Practical Telecommunications and Wireless Communications

For Business and Industry

Other titles in the series

Practical Data Acquisition for Instrumentation and Control Systems (John Park, Steve Mackay)

Practical Data Communications for Instrumentation and Control (Steve Mackay, Edwin Wright, John Park)

Practical Digital Signal Processing for Engineers and Technicians (Edmund Lai)

Practical Electrical Network Automation and Communication Systems (Cobus Strauss)

Practical Embedded Controllers (John Park)

Practical Fiber Optics (David Bailey, Edwin Wright)

Practical Industrial Data Networks: Design, Installation and Troubleshooting (Steve Mackay, Edwin Wright, John Park, Deon Reynders)

Practical Industrial Safety, Risk Assessment and Shutdown Systems for Instrumentation and Control (Dave Macdonald)

Practical Modern SCADA Protocols: DNP3, 60870.5 and Related Systems (Gordon Clarke, Deon Reynders)

Practical Radio Engineering and Telemetry for Industry (David Bailey)

Practical SCADA for Industry (David Bailey, Edwin Wright)

Practical TCP/IP and Ethernet Networking (Deon Reynders, Edwin Wright)

Practical Variable Speed Drives and Power Electronics (Malcolm Barnes)

Practical Centrifugal Pumps (Paresh Girdhar and Octo Moniz)

Practical Electrical Equipment and Installations in Hazardous Areas (Geoffrey Bottrill and G. Vijayaraghavan)

Practical E-Manufacturing and Supply Chain Management (Gerhard Greef and Ranjan Ghoshal)

Practical Grounding, Bonding, Shielding and Surge Protection (G. Vijayaraghavan, Mark Brown and Malcolm Barnes)

Practical Hazops, Trips and Alarms (David Macdonald)

Practical Industrial Data Communications: Best Practice Techniques (Deon Reynders, Steve Mackay and Edwin Wright)

Practical Machinery Safety (David Macdonald)

Practical Machinery Vibration Analysis and Predictive Maintenance (Cornelius Scheffer and Paresh Girdhar)

Practical Power Distribution for Industry (Jan de Kock and Cobus Strauss)

Practical Process Control for Engineers and Technicians (Wolfgang Altmann)

Practical Troubleshooting Electrical Equipment (Mark Brown, Jawahar Rawtani and Dinesh Patil)

Practical Power Systems Protection (Mark Brown and Ben Ramesh)

Practical Telecommunications and Wireless Communications

For Business and Industry

Edwin Wright BSc, BE (Hons), (ElecEng), MIPENZ,
IDC Technologies, Perth, Australia

Deon Reynders Pr. Eng, BSc (Hons), (ElecEng), MBA,
Senior Staff Engineer, IDC Technologies, Perth, Australia

Series editor: Steve Mackay FIE (Aust), CPEng, BSc (ElecEng), BSc (Hons),MBA,
Gov.Cert.Comp.,Technical Director – IDC Technologies

AMSTERDAM • BOSTON • HEIDELBERG • LONDON
NEW YORK • OXFORD • PARIS • SAN DIEGO
SAN FRANCISCO • SINGAPORE • SYDNEY • TOKYO

Newnes is an imprint of Elsevier

Newnes

Newnes
An imprint of Elsevier
Linacre House, Jordan Hill, Oxford OX2 8DP
30 Corporate Drive, Burlington, MA 01803

First published 2004

British Library Cataloguing in Publication Data
A catalogue record for this book is available from the British Library

Library of Congress Cataloguing in Publication Data
A catalogue record for this book is available from the Library of Congress

ISBN 0 7506 6271 9

For information on all Newnes publications
visit our website at www.newnespress.com

Typeset by Integra Software Services Pvt. Ltd, Pondicherry, India
www.integra-india.com

Printed and bound by CPI Group (UK) Ltd, Croydon, CR0 4YY
Transferred to Digital Print 2011

Contents

Preface

The make-up and structure of telecommunications networks has changed dramatically in the past few years. These changes impact on the equipment you purchase, the services you use, the providers you can choose and the means of transporting the data. This book will be of particular benefit to those who want to apply the latest and most effective telecommunications technology immediately. The book commences with a review of telecommunications basics and typical transmission media ranging from copper to fiber optics. This brings everyone up to speed with the fundamentals of telecommunications. Public Telephone network services are then examined with a review of the typical infrastructure (including ISDN and DDN). Narrowband networks are then investigated with a discussion on PABX's and Computer/Telephony integration. The topic of wide area networks and local area networks is then reviewed with a focus on the TCP/IP protocol as applied to telecommunications systems. All the important topics of Internet/IP applications and services is then analyzed with a discussion on topics such as Voice over IP. The meaning and structure of broadband communications is defined and the practical applications of such technologies as ADSL identified. Finally, the fast expanding topic of wireless communications is investigated in considerable depth with a focus on practical applications and breaking developments, which you can use in your work.

Throughout the book you will receive guidance and practical tips from a proven expert about how to apply this technology to your company. Your company may already be looking at operating its own telecommunications system or may be looking at using the systems on the market. With the vast array of equipment and systems and technology now available to you, you need the necessary knowledge to make the best decisions. We believe this book will allow you to achieve your objectives in learning and then applying the fundamentals of telecommunications to your next project.

This book has been designed for those who require a basic but fundamental grounding in telecommunications and will be of special benefit for those who want to apply the technology as quickly as possible.

We would hope that you will gain the following knowledge from this book:

- The fundamentals of telecommunications
- The 'jargon' used in telecommunications
- The 'nuts and bolts' about selecting and installing telecommunications systems
- How to increase the bandwidth by exploiting your existing copper wire more effectively
- How to make 'best practice' decisions on the best and most cost effective access options for your company
- How to apply the latest technologies such as wireless communications
- How to understand and apply high speed access systems such as ADSL and beyond.

Typical people who will find this book useful include:

- Electrical Engineers
- Technicians
- Managers
- Instrumentation Engineers & Technicians
- Process Control Engineers & Technicians

- Project Engineers
- Systems Engineers
- Process Engineers
- Maintenance Engineers
- Sales Engineers
- Engineering Managers
- Network Administrators
- Software Engineers
- Field Technical Support Staff.

A modicum of electrical and data communications knowledge will enable you to maximize your understanding of the material in the book.

1

Introduction to telecommunications

1.1 Telecommunications

The word 'telecommunication' is derived from the Greek stem 'tele' meaning 'at a distance' and the word 'communications' meaning 'the science and practice of transmitting information'. A more useful definition is given in the Dictionary of Communications Technology as 'a term encompassing the transmission or reception of signals, images, sounds, or information by wire, radio, optic, or infra-red media'.

Telecommunications plays a vital role in international commerce, and in industrialized nations it is an accepted necessity. The telecommunications networks in all countries are linked together to form a global telecommunications network for carrying information of all kinds. The public switched telephone network (PSTN) was originally developed solely for carrying voice communications, but today carries ever-increasing data communication traffic. The Internet uses the PSTN circuits to carry its data, and the phenomenal growth of the Internet has stimulated the growth of data circuit usage in the PSTN. In some countries, Internet traffic accounts for more than half the total PSTN traffic.

Wireless services are having an enormous impact on the growth of telecommunications networks. In industrialized countries they are used increasingly for mobile business communications. On the other hand, in developing countries, they enable many customers in the main population centers to obtain affordable telecommunications services. The telecommunications provider does not have to invest in the very high costs of fixed subscriber distribution plant for the individual customers and can serve thousands of customers from one transmitter site. Service can be supplied almost immediately – in the time taken to purchase a cellular phone, sign a service contract and arrange the network connection by means of a telephone call or data message to the service provider.

1.2 Principles of telecommunications services

Telecommunications services follow these principles:

- The telecommunications networks are used to provide services to the users.
- A service requires the execution of a series of programs by the originating and destination entities.
- The services are decomposed into different layers by the initiating entity, where each layer undertakes a specified portion of the overall service. This makes the services more manageable and allows interoperability between vendors. The OSI reference model explains this layered architecture.

- The telecommunications services include information transfer, signaling, and billing.
- Information is transferred over the network in the form of bits. These bits have different forms depending on the type of the transmission media; electrical signals on copper cables, pulses of light in fiber-optic cables and electromagnetic waves traveling through space in radio signals.
- Signals can be corrupted during transmission due to interference.
- Protocols incorporate error correction and detection mechanisms to overcome errors.

1.3   Chapter overview

Chapter 2 is to familiarize the reader with some of the basic telecommunications concepts. In this chapter the following areas will be discussed; types of channels and their methods of operation, modulation methods as ways of imposing information on the channel, the difference between analog and digital channels and different ways of making connections across the network. The open systems interconnection (OSI) model introduces the concept of layered architectures and the functions of the various layers are explained. More experienced readers may wish to skip this chapter.

Chapter 3 examines some of the different types of transmission media used for physically conveying signals from one point to another. The approach taken will be to explain the fundamental method of operation of each of these transmission media types, introduce the various system components and discuss the application for each type. Some of the main bearer design considerations will be discussed to enable the reader to make an informed decision as to which type of media to use for a particular application. The discussion will commence with systems guided over a physical bearer; namely twisted pair and coaxial copper, fiber-optic cables and the power distribution system. The discussion will then move on to wireless systems; namely microwave radio systems, satellite systems and infra-red transmission, which require no specific bearer and radiate their signals as electromagnetic waves. The emphasis in this chapter is on the fundamental transmission bearer systems, Chapter 6 discusses the methods of carrying information, both analog and digital, on these bearer systems.

Chapter 4 discusses the structure of the PSTN, together with the CCITT signaling system No. 7, which provides a common signaling channel across the digital networks to enable many sophisticated network services, such as ISDN, to be provided.

Chapter 5 discusses various aspects of the voice switching equipment located in a customer's premises. This includes PBX systems, key systems and centrex service. The discussion covers computer telephony integration (CTI), which allows a computer and a telephone to work together so that a user can manage their telephony services using a PC in conjunction with a telephone. For example this allows customer details to pop up on the computer screen when the call is answered.

Chapter 6 considers the different communication approaches used for the provision of services in the public telecommunications network. Fundamentally circuits can be divided into four categories: either analog or digital and each of these can be either switched or leased. First we will consider the switched analog network and then look at the use of dedicated or leased analog circuits. Next the digital services will be discussed beginning with the various types of switched digital services and concluding with various alternatives for dedicated digital services, primarily directed at the large users. In each section the method of operation will be explained, the circuit characteristics described and the particular advantages the method offers will be considered. This is intended to enable the reader to make an informed choice about the appropriateness of that particular approach for their unique situation.

Chapter 7 examines the available technologies for the provision of broadband customer access at greater than 1 Mbps, including both wired and wireless services. The wired systems discussed include: systems operating over the existing POTS copper pairs, known collectively as xDSL systems, systems used to provide existing cable television services such as hybrid fiber coax (HFC) and their new counterparts such as fiber to the curb (FTTC) and fiber to the home (FTTH). The wireless systems include multi-channel multipoint distribution system (MMDS), local multipoint distribution services (LMDS), and short-range systems such as Bluetooth. This chapter discusses the operation of each of these alternative customer access technologies together with their relative merits and associated performance issues.

Chapter 8 looks at the following aspects of providing local area networks to business customers:

- The various LAN topologies as well as their advantages and disadvantages
- Various media access methods, specifically CSMA/CD, CSMA/CA, token passing and polling
- Various LAN standards with emphasis on IEEE 802.3, IEEE 802.5, IEEE 802.11 and ANSI X3T9.5
- The functionality devices used to interconnect networks: repeaters, bridges, routers, switches and gateways
- The basic characteristics of LANs, WANs, MANs, VLANs and VPNs.

Chapter 9 deals with the convergence of conventional PSTN networks and IP based internetworks, in particular the Internet. As a result of this convergence, 'voice over IP' is making major inroads into the telecommunications industry. This chapter will introduce the ITU-T H.323 standard for multimedia (audio, video and data) transmission and discuss the TCP/IP protocol suite sufficient for the understanding of the operation of this protocol.

Chapter 10 explains the basic operation of cellular wireless communication systems. The following wireless technologies are discussed:

- Analog cellular voice systems, in particular AMPS and N-AMPS
- Digital cellular voice systems, in particular D-AMPS (North American TDMA), CDMA and GSM
- Cellular data systems, in particular CDPD, GPRS and HDR
- 'Cordless' technologies, in particular CTS and DECT
- Personal communications service (PCS)
- Wireless applications protocol (WAP)
- Third generation (3G) mobile technologies.

Chapter 11 provides in-depth information about wireless LAN technologies. The following areas are discussed:

- IEEE 802.11, 802.11b, 802.11a
- Wireless LAN security issues
- Wireless personal area networks (WPAN)
- Blue tooth/IEEE 802.15.

1.4   Telecommunications standards

Telecommunications standards are essential to enable the global PSTN to function. It is clear that when a telephone call or data message originates in one country and terminates in another, both the sender and recipient need to understand each other's messages. This is most easily achieved by using standardized message formats. This enables a message to be successfully passed through a number of countries, as necessary, along the way.

This standardization is provided by two international organizations; the International Telecommunication Union (ITU), and the International Standardization Organisation (ISO). These are supplemented by many national standardization agencies.

The ITU provide recommendations, which serve as worldwide standards, although they are not legally binding. Prior to January 1993, their telecommunication recommendations were produced by the CCITT, the International Consultative Committee for Telephone and Telegraph, which has now been reorganized into the telecommunication standardization sector of the ITU. Their recommendations are denoted ITU-T. Similarly the International Consultative Committee for Radio (CCIR) handled the standardization of radio communications. This has now become the ITU radiocommunication sector and provides ITU-R recommendations.

The ISO has issued many important data communications standards. One of the most important is the open systems interconnection reference model, which we shall discuss later. Many national standards organizations are affiliated to ISO, including the American National Standards Institute (ANSI).

Some of the other important telecommunication standards organizations are:

- Electronics Industries Association (EIA)
- Telecommunication Industry Association (TIA)
- European Telecommunication Standardisation Institute (ETSI)
- Institute of Electrical and Electronic Engineers (IEEE)
- International Electrotechnical Commission (IEC).

Telecommunications standards are also developed by groups of manufacturers and users who formulate standards. These become ad hoc industry standards and may subsequently be incorporated in the recommendations of the international standards organizations. Some examples of these groups are the frame relay forum, the ADSL forum and the ATM forum.

2

Telecommunication basics

Objectives

The purpose of this chapter is to familiarize the reader with some of the basic telecommunications concepts used in this book. Where applicable, the user will be referred to the chapter dealing with that specific topic in greater depth.

2.1   Concepts

2.1.1   Signaling

In order to place a telephone call successfully, a signaling protocol has to convey information about the call through the telephone system in order to control, route and maintain the call. This information includes, for example, the number being called and signals representing specific conditions such as line free or subscriber unavailable. There are three different types of signals, described in detail in Chapter 4:

1. Addressing signals that represent the number being called
2. Alerting signals, indicating to the end-user that a call has arrived, and
3. Supervisory signals, indicating to the caller whether the line is free, the phone is ringing on the other side, etc., and to the exchange whether the caller's handset is off-hook or on-hook.

On older systems (prior to the mid-70s) this signaling was done by means of voltage levels and pulses or tones that represent the necessary information, using the same channel as the actual voice communication. This is referred to as in-band signaling, a process that is slow and wastes a significant amount of network capacity.

Because of the limitations of in-band signaling, common channel signaling (CCS) has been introduced. Here the signaling takes place on a separate signaling network. The current CCS ITU standard is Signaling System 7 (SS#7) and is described in Section 4.8.

2.1.2   Circuit

In telecommunications, a circuit is a physical electronic path that carries electronic information, be that voice or data, either in digital or analog format, between two points.

2.1.3   Channel

A channel refers to a 'logical' transmission path. For example, a particular radio station is allocated an FM channel centered on a specific frequency of 94.5 MHz. Using multiplexing techniques such as frequency division multiplexing (FDM), several transmission paths i.e. channels (based on different frequencies) can be created across a single medium and allocated to different users.

2.1.4   Line

A line is a telephone connection between a user and an exchange point set up by a telecommunications carrier.

2.1.5   Trunk

In telecommunications, the cable group that forms the primary path between two switching stations is known as a trunk. As such, it handles large volumes of traffic.

2.1.6   Bandwidth

The quantity of information a channel can convey in a given period is determined by its ability to handle the rate of change of the signal, i.e. its frequency. The frequency of an analog signal varies between a minimum and a maximum value and the difference between those two frequencies is the bandwidth of that signal.

The bandwidth of an analog channel is the difference between the highest and lowest frequencies that can be reliably transmitted over the channel. Bandwidth is normally specified in terms of those frequencies at which the signal has fallen to half the power (or 0.707 of the voltage) relative to the mid-band frequencies, referred to as –3dB points. In this case the bandwidth is known as the –3dB bandwidth (Figure 2.1).

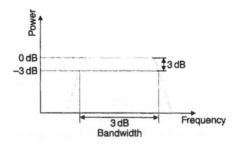

Figure 2.1
Channel bandwidth

Digital signals are made up of a large number of frequency components, but only those within the bandwidth of the channel will be able to be received. It follows that the larger the bandwidth of the channel, the higher the data transfer rate can be and more high frequency components of the digital signal can be transported, and so a more accurate reproduction of the transmitted signal can be received (Figure 2.2).

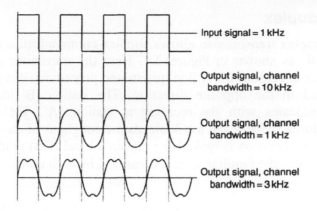

Input signal = 1 kHz

Output signal, channel bandwidth = 10 kHz

Output signal, channel bandwidth = 1 kHz

Output signal, channel bandwidth = 3 kHz

Figure 2.2
Effect of channel bandwidth on a digital signal

2.1.7   Channel capacity

The channel capacity i.e. the maximum data transfer rate of the transmission channel can be determined from its bandwidth, by using the following formula derived by Shannon.

$$C = 2B \log_2 M \text{ bps}$$

where C is the channel capacity, B the bandwidth of the channel in Hertz and M the discrete levels used for each signaling element.

In the special case where only two levels, 'ON' and 'OFF' or 'HIGH' and 'LOW' are used (binary), $M = 2$. Thus $C = 2B \log_2 2$ but $\log_2 2 = 1$, therefore $C = 2B$. As an example, the maximum data transfer rate for a PSTN channel of bandwidth 3200 Hz carrying a binary signal would theoretically be $2 \times 3200 = 6400$ bps. In practice this figure is largely reduced by other factors such as the presence of noise on the channel to approximately 4800 bps.

2.2   Simplex, half-duplex and full-duplex

2.2.1   Simplex

A simplex channel is unidirectional and allows data to flow in one direction only, as shown in Figure 2.3. Public radio broadcasting is an example of a simplex transmission. The radio station transmits the broadcast program, but does not receive any signals back from the radio receiver.

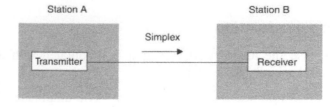

Station A

Station B

Simplex

Transmitter

Receiver

Figure 2.3
Simplex transmission

This has limited use for data transfer purposes, as invariably the flow of data is required in both directions in order to control the transfer process, acknowledge data, etc.

2.2.2   Half-duplex

Half-duplex transmission allows simplex communication in both directions over a single channel, as shown in Figure 2.4. Here the transmitter at station 'A' sends data to a receiver at station 'B'. A line turnaround procedure takes place whenever transmission is required in the opposite direction. The station 'B' transmitter is then enabled and communicates with the receiver at station 'A'. The delay in the line turnaround procedures reduces the available data throughput of the communications channel. This mode of operation is typical for citizen's band (CB) or marine/aviation VHF radios and necessitates the familiar 'over!' command by both users.

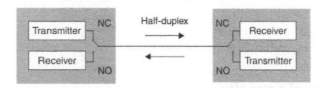

Figure 2.4
Half-duplex transmission

2.2.3   Full-duplex

A full-duplex channel gives simultaneous communications in both directions, as shown in Figure 2.5.

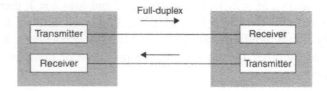

Figure 2.5
Full-duplex transmission

2.3   Modulation techniques

Modulation can involve either analog or digital signals, or both. It refers to the process of superimposing the information (modulating) signal on to a carrier signal. In its unmodulated state, the carrier is a constant amplitude, constant frequency signal.

2.3.1   Analog modulation

In this case the analog carrier signal is modulated with an analog information signal. There are basically three methods:

1. *Amplitude modulation (AM)*:   The amplitude of the carrier is varied by the amplitude of the information signal.
2. *Frequency modulation (FM)*:   The frequency of the carrier is varied by the amplitude of the information signal.
3. *Phase modulation (PM)*:   The phase (time displacement) of the carrier signal is varied by the amplitude of the information signal.

2.3.2 RF modulation

RF modulation is similar to analog modulation, with the exception that the input (modulating) signal is digital.

- *Amplitude shift keying (ASK)*: The amplitude of the carrier signal is varied between two values. This is also known as *on–off keying (OOK)*.
- *Frequency shift keying (FSK)*: The frequency of the carrier signal is varied between two values by the modulating signal.
- *Phase shift keying (PSK) a.k.a. binary phase shift keying (BPSK)*: The phase of the carrier signal is changed by the modulating signal. Depending on the number of discrete displacements, several bits of data can be transmitted simultaneously. For example, with four shift amounts (0°, 90°, 180° and 270°), two bits e.g. 00, 01,10 and 11 can be sent at a time.
- *Quadrature amplitude modulation (QAM)*: Both the phase *and* the amplitude of the carrier are changed, making it possible to encode as many as 4 bits at a time.
- *Trellis coded modulation (TCM)*: This is similar to QAM, but includes extra bits for error correction.

The shift keying modulation methods come in plain and differential forms. The differential versions encode values as changes in a parameter, not in a specific value for a parameter. The differential techniques are easier to implement, and more robust than the non-differential ones.

- *Differential amplitude shift keying (DASK)*: Similar to ASK, but encoding different digital values as changes in signal amplitude.
- *Differential frequency shift keying (DFSK)*: Similar to FSK, but encoding different values as changes in signal frequency.
- *Differential phase shift keying*: Similar to PSK, but encoding different digital values as changes in signal phase.

2.3.3 Digital modulation

Digital modulation converts an analog signal into a series of binary digits for subsequent transmission. All these methods sample the analog input signal at a pre-defined rate and then generate a binary output based on that sample.

- *Delta modulation (DM)*: The analog signal is represented by a series of bits (1's and 0's) that represent the current amplitude of the input signal relative to the previous amplitude. If the signal increases, then 1's are sent. If the signal decreases, 0's are sent. If the signal remains constant, alternating 1's and 0's are sent. Variations on the theme include adaptive delta modulation (ADM).
- *Pulse code modulation (PCM)*: It is typically used to convey voice signals across a digital channel. The analog signal sample is converted into an n-bit digital number that is subsequently transmitted. A 7 bit code means that the value transmitted will represent the analog voltage sample taken to within $1/(2^7) = 1/128$th of its original value. This is sufficient for voice applications, but for multimedia applications up to 24 bits may be needed. The transmission rate is typically 64 kbps. Variations on the theme include adaptive differential pulse code modulation.

- *Pulse width modulation (PWM) a.k.a. pulse duration modulation (PDM)*:   An analog value is represented by changing the width (duration) of a discrete pulse.
- *Pulse amplitude modulation (PAM)*:   An analog value is represented by the amplitude of the carrier for that interval.
- *Pulse position modulation (PPM)*:   An analog signal is represented by varying the position (i.e. the displacement) of a discrete pulse within a bit interval.

2.4   Baseband vs broadband

2.4.1   Baseband

Baseband refers to a communication method where all the traffic shares a single channel; hence when a given user transmits on the channel, the entire bandwidth is occupied and nobody else can use the channel. If more than one user wish to use the channel, they need to do it sequentially; i.e. they have to use a time-related multiplexing scheme such as TDM.

The signal is placed on the medium without using a high-frequency carrier signal (Figure 2.6). In the case of analog signals, such as on an ordinary telephone, the voice signal is simply transmitted 'as is'; i.e. no modulation or encoding technique is used. In the case of digital data, such as RS-232 data, the 0's and 1's are simply encoded as a series of positive and negative voltages.

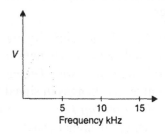

Figure 2.6
Baseband signal

A benefit of baseband is the simplicity of the system. No complex modulation and demodulation equipment is required. On the downside, the signal at the receiving end has to be large enough to be detected by the line receiver. For a digital system operating at, say, 10 Mbps, this may limit the distance to a few hundred meters. Another drawback is that only one signal can be transmitted at a given time.

2.4.2   Broadband

Broadband refers to a system where multiple signals are transmitted simultaneously over a common physical channel (Figure 2.7). This is accomplished by using frequency division multiplexing (for systems transmitting data using copper or radio) or wave division multiplexing (for systems transmitting data using fiber optics), and transmitting different channels at different frequencies or wavelengths.

The various input signals are used to modulate their allocated carrier signals, using for example FM or AM, and these modulated carriers are combined and sent across the medium as one composite signal. Even if the input is a digital signal, the result is still a modulated analog signal. At the receiving end the various signal components are extracted via band-pass filters, and the information is recovered via appropriate demodulators or discriminators.

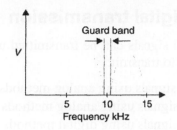

Figure 2.7
Broadband signal

The advantage of broadband transmission is that several channels of information can be sent concurrently over one physical medium. Because analog carrier signals (sine waves) are employed, it is also possible to recover the original signal from a very small received signal, even down to the picovolt range with appropriate technology. Hence broadband can be used across very large distances, in fact across millions of kilometers in the case of discovery satellites.

On the downside, the transmission and reception circuitry is more complex and hence more expensive than in the case of baseband.

2.5 Narrowband vs wideband

2.5.1 Narrowband

Although narrowband and wideband sound very much like baseband and broadband, they have absolutely nothing in common. Narrowband and wideband refer to the relative bandwidth of the channel.

A narrowband system is simply a system with a relatively small bandwidth. The following systems can be classified as narrowband:

- Links designed to connect teletypes with a bandwidth of 300 Hz.
- A plain old telephone system (POTS) with an available bandwidth of about 3 kHz. With the aid of modems they can carry data at speeds up to 56 kbps.
- Narrow band ISDN (Integrated Services Digital Network) with two channels for voice or data, at 64 kbps each.
- A digital voice channel with a bandwidth of between 64 kbps and 1.54 Mbps.

2.5.2 Wideband

The following systems can be classified as wideband:

- T-3 operating at 44.7 Mbps. It is equivalent to 28 T-1 circuits and can transmit 672 conversations simultaneously over fiber optics or microwave.
- ATM, carrying data, voice and video at speeds up to 13.22 Gbps.
- Likewise SONET, an optical multiplexing interface for high-speed data transmission, can operate up to 13.22 Gbps.
- Cable TV (CATV), broadcasting local and satellite TV channels using 500 MHz bandwidth.
- Digital high-definition TV (HDTV), used for broadcasting TV at 6 MHz per channel.

2.6   Analog vs digital transmission

Analog and digital signals can be transmitted using either analog or digital methods. It is therefore possible to transmit:

- Analog signals using analog methods
- Digital signals using analog methods
- Digital signals using digital methods
- Analog signals using digital methods.

Analog signals may change continuously in both frequency and amplitude. These signals are commonly used for audio and video communication as illustrated in Figure 2.8.

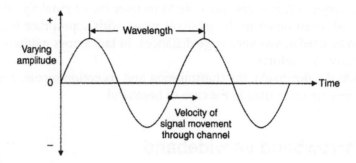

Figure 2.8
Analog signal

Digital signals, on the other hand, are characterized by the use of discrete signal amplitudes (Figure 2.9). A binary digital signal, for example, has only two allowed values representing the binary digits 'ON' and 'OFF'. In fiber-optic communications channels the presence or absence of light normally represents these states.

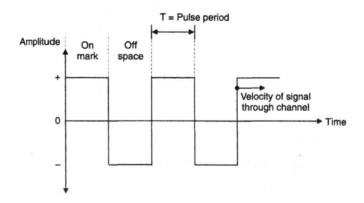

Figure 2.9
Digital signal

2.6.1   Analog signals over analog channels

This is typically the case with voice over a POTS telephone line. For this purpose a normal dial-up or leased analog line is used as outlined in Chapter 6. No sophisticated conversion or modulation techniques are necessary.

2.6.2 Digital signals via analog methods

In this case, the method used to transmit the signal involves a modulated carrier signal. A sine wave is modulated with the digital data and, if necessary, several channels can be superimposed on the same physical circuit. Commercial solutions to this problem include modems for use on phone lines or one of the broadband alternatives outlined in Chapter 7.

2.6.3 Digital signals via digital methods

No special conversion techniques are needed. The transmission channel could be any of the dial-up or leased alternatives such as T1 or ATM as described in Chapter 6.

2.6.4 Analog signals via digital methods

Here the channel expects digital information and for this reason the analog signal first has to be converted to a digital format. A typical solution is pulse code modulation (PCM) as covered elsewhere in this chapter. In the case of voice, the analog signal is typically sampled at 8000 times per second and each sample is then converted to an 8 bit binary number giving a 64 kbps data rate.

2.7 Dial-up vs leased access

2.7.1 Dial-up access

Dial-up access refers to connecting a device to a network via a public switched telephone network and a modem. The procedure is the same as for two people establishing contact via telephone, the difference being that two computer devices are communicating via modems.

The connections are not always good, causing modems to drop out or fail to establish connections, and the data transfer rates are limited to 56 kbps or less, depending on the maximum frequency that can be negotiated by the modems. Provided that multiple connections are supported at both the user end and the service provider end (e.g. through multilink PPP), multiple parallel dial-up connections can be established for high volumes of data.

An alternative to a normal telephone connection is an integrated services data network (ISDN) connection. One ISDN BRI (basic rate interface) can handle 128 kbps of data, and if the necessary services are installed, multiple ISDN connections can be established to handle high data volumes, e.g. in the case of videoconferencing.

2.7.2 Leased access

A leased line (also referred to as a dedicated or a private line) is a permanent line installed between locations, usually to a user's premises, by a telephone company. Unlike a dial-up connection it is always active. Leased lines are available in 2 and 4 wire versions, which has a bearing on the line quality and obviously also on the line rental. Leased lines provide faster throughput of data and better quality connections than dial-up, but at a higher cost.

Examples of leased line services are:

- *T-1/E-1 lines*: These provide 1.544 Mbps for T-1 (e.g. in the United States and Japan) or 2.048 Mbps for E-1 in Europe and Mexico).
- *Fractional T-1 lines*: These build up in units of 64 kbps, up to 768 kbps.

- *56/64 kbps lines*: In Europe, these lines provide 64 kbps, In the US and other countries 8 kbps is used for control overhead so that only 56 kbps is available to the user.
- *Digital data services (DDS)*: These provide synchronous transmission of digital signals at 2.4, 4.8, 9.6, 19.2 or 56 kbps.

The line is rented on a monthly basis and the rate is affected by factors such as the speed of the circuit and the distance involved. The availability and pricing varies between different service providers.

Ordinary telephone lines were primarily intended for voice traffic and therefore their performance for data transmission is far from optimum. At a cost, specially conditioned analog lines can be leased in order to overcome this problem. Leased line services are discused in Chapter 6.

2.8   Multiplexing techniques

Multiplexing is a technique by which information (voice or data) from more than one source is delivered across one common medium, and delivered to the appropriate recipient. This section describes various methods to accomplish this.

2.8.1   Space division multiplexing (SDM)

Space division switching involves the physical connection of one path to another. This kind of switching was used in all the earliest exchanges, and is still an essential part of modern digital exchanges.

2.8.2   Time division multiplexing (TDM)

In TDM, multiplexing takes place in the time domain (Figure 2.10). Data streams from several sources are combined by sending small slices of data from each input across the common channel sequentially, in time, in a specific sequence. If there are 4 channels to be multiplexed like this, each source will have access to the channel for 1/4 (25%) of the time.

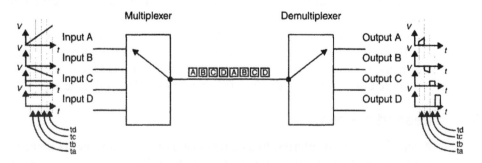

Figure 2.10
TDM

Typical examples of TDM are T-1 and T-3 carrier systems, where PCM streams are created for each conversation or data stream, and then combined with TDM on to a common channel.

A variation on the theme is statistical multiplexing (STDM), where the input nodes that have nothing to send are skipped, thereby increasing the available time for the remaining nodes.

2.8.3 FDM

In frequency division multiplexing (FDM), multiplexing takes place in the frequency domain (Figure 2.11). The various signals are separated not in time, but in frequency. FDM assigns a discrete (fixed) carrier frequency for each channel, and then modulates each carrier with the input for the respective channel. Assuming that there are 4 channels and the total available bandwidth is between 1 and 2 MHz, the maximum available bandwidth per channel is $(2 - 1)/4 = 1/4$ MHz or 250 kHz. In practice this will be slightly less due to a 'guard band' separating the channels on the frequency domain.

Typical examples of FDM applications are commercial TV and radio broadcasts.

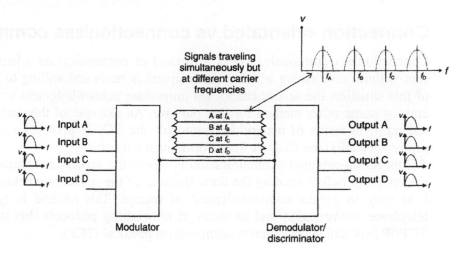

Figure 2.11
FDM

2.8.4 WDM

Wavelength division multiplexing (WDM) is used on optical fibers and is the optical equivalent of FDM (Figure 2.12). Light waves with different light wavelengths (i.e. different 'colors') are used to carry separate streams of information, thereby creating several channels on the same optical fiber. The different modulated light beams are combined and transmitted as one composite light beam down the fiber, and at the receiving end it is optically split into its individual streams, which are subsequently demodulated.

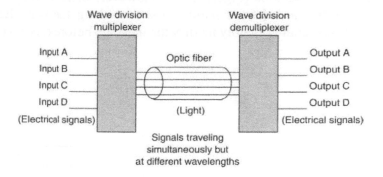

Figure 2.12
WDM

2.8.5   DWDM

Dense waveform division multiplexing (DWDM) is related to WDM and offers even higher bandwidths.

It effectively creates several virtual fibers in one physical fiber, each virtual fiber carrying several channels. Thus, for example, eight optical channels of 2.5 Mbps each can be combined to give an effective data rate of 20 Mbps. The technology involved is developing rapidly and by the end of 1999 it was already possible to carry 96 seperate data streams.

With a new technology called UDWDM (ultra-dense WDM) Bell Labs have prototyped a system that can run 1000 channels over a single fiber, each channel operating at up to 160 Gbps.

2.9   Connection orientated vs connectionless communication

Connectionless communication is a method of communication whereby data is simply sent without confirming whether the recipient is ready and willing to receive it. Because of this situation the sender cannot get immediate acknowledgment of receipt, and has to employ some other method for this purpose. An example of this technique is sending a telegram. In terms of networking protocols the TCP/IP protocol suite's user datagram protocol (UDP) (see Chapter 9) performs such a function.

Connection-oriented communication is where the sender first establishes contact with the recipient before sending the data. Because of the connection between the two parties, it is easy to obtain acknowledgment of receipt. This method is typical of a normal telephone conversation and in terms of networking protocols this is performed by the TCP/IP protocol suite's transmission control protocol (TCP).

2.10   Types of transmission

The following transmission types are used for either analog or digital transmissions, or for both.

2.10.1   Point-to-point

Point-to-point transmission takes place between a single pair of stations, normally relatively close to each other. An example of this is a home intercom system with two stations.

2.10.2   Mediated

A message is sent from one station to another, but because of the distance there are intermediate stations (Figure 2.13). However, there is only one possible path and the signal is forwarded from station to station along the way. Each station along the way handles and could possibly modify the message before it is re-transmitted.

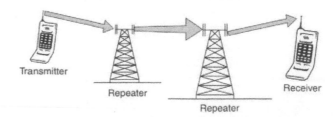

Figure 2.13
Mediated transmission

2.10.3   Switched

This is the same as mediated transmission, with the difference being that there are multiple paths between source and destination. The nodes in between can therefore switch (i.e. divert) incoming traffic and send it on to an appropriate node. The word 'switch' in this context has the same implication as a 'switch' (a set of 'points') on a railroad track.

There are essentially two basic modes of switching, namely circuit switching and packet switching.

2.10.4   Circuit switching

In a circuit switching system a continuous connection is made across the network between the two end points (Figure 2.14). This is a temporary connection which remains in place as long as both parties wish to communicate, that is until the connection is terminated. All the network resources are available for the exclusive use of these two parties whether they are sending data or not. When the connection is terminated the network resources are released for other users.

The advantage of circuit switching is that the users have an exclusive channel available for the transfer of their data at all times while the connection is made. The obvious disadvantage is the cost of maintaining the connection when there is little or no data being transferred. Such connections can be very inefficient for the bursts of data that are typical of many computer applications.

A telephone call is a good example of a circuit switching system.

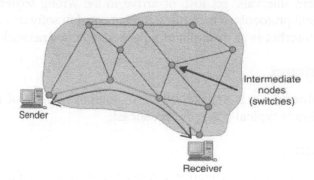

Figure 2.14
Circuit switching

2.10.5   Packet switching

Packet switching systems improve the efficiency of the transfer of bursts of data, by sharing communications channels among many similar users (Figure 2.15). This is analogous to the postal system. The unit of information may vary. It could be an entire message, or a block of data with a small part of a long message.

Packet switched messages are broken into a series of packets of a certain maximum size, each containing the destination and source addresses and a packet sequence number. The packets are sent over a common communications channel, possibly interleaved with those of other users. All the receivers on the channel check the destination addresses of all packets and accept only those carrying their address. Messages sent in multiple packets are reassembled in the correct order by the destination node.

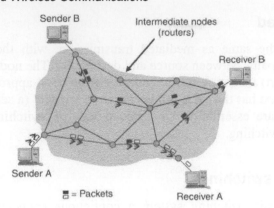

Figure 2.15
Packet switching

All packets do not necessarily follow the same path. As they travel through the network they may get separated and handled independently of each other, but eventually arrive at their correct destination. For this reason, packets often arrive at the destination node out of their transmitted sequence. Some packets may even be held up temporarily (stored) at a node, due to unavailable lines or technical problems that might arise on the network. When the time is right, the node allows the packet to pass or be 'forwarded'.

This method of delivery is obviously more cost efficient than circuit switching since data from any user can travel over any portion of the system at any time, leading to more efficient use of the available infrastructure. On the downside, packets may be delayed, arrive at non-consistent intervals, get lost, or arrive in the wrong sequence. This necessitates the use of additional protocols (sets of rules, implemented in software) to take care of the problem.

The Internet is an example of a packet-switched network.

2.10.6   Broadcast

The information is transmitted to all stations capable of receiving, and not to a specific one. This is typical of a radio broadcast.

2.10.7   Multicast

The information is selectively transmitted to a specific sub-group of devices, within a larger group of devices all capable of receiving the same signal. An example is 'pay TV', either cable or satellite based, where all receivers pick up the same signal but only those with a decoder can display it.

2.10.8   Stored and forwarded

Data is sent to a holding location until requested or sent on automatically after a predefined amount of time.

2.11   Local vs wide area networks

2.11.1   Local area networks

Local area networks (LAN) are characterized by high-speed transmission over a relatively restricted geographical area such as a building or a group of buildings. LANs consist of a common medium (such as a coaxial cable), interconnecting computers, printers, programmable logic controllers, etc. Users can share resources such as printers, transfer data to each other, and communicate via e-mail or chat sessions.

There are many types of LANs, characterized by their topologies (the geometric arrangement of devices on the network), protocols (the rules for sending and receiving the data), media (e.g. the cable used) and media access methods (used to control the access of individual nodes to the medium). These will be discussed in more detail in Chapter 8.

A typical LAN is Ethernet 10BaseT, operating at 10 Mbps over a maximum distance of 100 m across unshielded twisted-pair wire via a hub. This concept is illustrated in Figure 2.16.

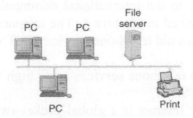

Figure 2.16
Local area network

2.11.2   Wide area networks

While LANs operate where distances are relatively small, wide area networks (WANs) consist of inter-linked LANs that are separated by large distances ranging from a few tens of kilometers to thousands of kilometers. WANs normally use the public telecommunications carriers to provide cost-effective connection between LANs. Since these links are supplied by independent telecommunications utilities, they are commonly referred to (and illustrated as) a 'communications cloud'. Special equipment called gateways (routers) have been developed for this type of activity, storing the message at LAN speed and transmitting it across the 'communications cloud' at the speed of the interconnecting carrier. When the entire message has been received at the remote LAN, the message is reinserted at the local LAN speed. The speed at which a WAN interconnects is often slower than the LAN speed, but not necessarily so since many of the WAN carrier technologies such as ATM are capable of speeds far in excess of typical LAN speeds. At the end of the day it boils down to cost.

The concept of a WAN is shown in Figure 2.17.

Figure 2.17
Wide area network

2.11.3   The 'communications cloud'

The various long-distance WAN transport services such as T1, E1, SONET, ATM, etc., are normally drawn as a 'communications cloud' with the end users on the inside and the carriers on the inside. This is done for a reason. The users' systems are normally obscured to the carriers, and the detail of the carriers are normally obscured to the users, hence the 'cloud'.

The carriers are often extended to the premises of the client. In the case of a T-carrier, it is terminated in a channel service unit or CSU on each side, installed on the user's premises. Between the CSUs the signal is the utility's problem, beyond the CSUs on both sides it is the user's problem.

2.12   The PSTN vs the Internet

The PSTN refers to the international communications infrastructure for carrying voice and data as described in Chapter 4. The telephone service provided by the PSTN is also referred to as plain old telephone service (POTS). POTS are normally restricted to about 56 kbps.

The high-speed telephone services using high-speed digital lines, such as ISDN, are not classified as POTS.

In contrast, the Internet is a global packet-switching network, originally designed for the transportation of digital data. This digital data is carried on the digital PSTN circuits described in Chapter 6. Almost all Internet users have access to the PSTN as well, and in fact most private users use the PSTN to gain access to their Internet service providers. Because of this co-existence, the idea of using the Internet to carry traditional PSTN information such as voice and fax (in digital format) has emerged, and a very significant current development is that of the 'converged' network as discussed in Chapter 9.

2.13   The open systems interconnection model

2.13.1   Overview

A communication framework that has had a tremendous impact on the design of communications systems is the open systems interconnection (OSI) model developed by the International Standardization Organization. The objective of the model is to provide a framework for the coordination of standards development and allow both existing and evolving standards activities to be set within that common framework.

The interconnection of two or more devices with digital communication is the first step towards establishing a network. In addition to the hardware requirements, the software problems of communication must also be overcome. Where all the devices on a network are from the same manufacturer, the hardware and software problems are usually easily solved because the system is usually designed within the same guidelines and specifications.

When devices from several manufacturers are used on the same application, the problems seem to multiply. Systems that are specific to one manufacturer and which work with specific hardware connections and protocols are called closed systems. Usually, these systems were developed at a time before standardization or when it was considered unlikely that equipment from other manufacturers would be included in the network.

In contrast, open systems are those that conform to specifications and guidelines which are 'open' to all. This allows equipment from any manufacturer, who complies with that standard, to be used interchangeably on the network. The benefits of open systems include multiple vendors and hence wider availability of equipment, lower prices and easier integration with other components.

In 1978 the ISO, faced with the proliferation of closed systems, defined a 'Reference Model for Communication between Open Systems' (ISO 7498), which has become known as the Open Systems Interconnection model, or simply as the OSI model. OSI is essentially a data communications management structure, which breaks data communications down into a manageable hierarchy of seven layers. Each layer has a defined purpose and interfaces with the layers above it and below it. By laying down standards for each layer, some flexibility is allowed so that the system designers can develop protocols for each layer independent of each other. By conforming to the OSI standards, a system is able to communicate with any other compliant system, anywhere in the world.

It should be realized at the outset that the OSI reference model is not a protocol or set of rules for how a protocol should be written but rather an overall framework in which to define protocols. The OSI model framework specifically and clearly defines the functions or services that have to be provided at each of the seven layers (or levels).

Since there must be at least two sites to communicate, each layer also appears to converse 'horizontally' with its peer layer at the other end of the communication channel in a virtual (logical) communication. The OSI layering concept is shown in Figure 2.18.

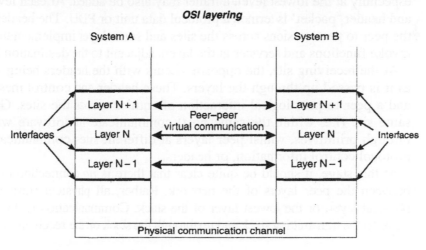

Figure 2.18
OSI layering concept

The actual functions within each layer are provided by entities such as programs, functions, or protocols, and the services for a particular layer are implemented on a single machine. Several entities, for example a protocol entity and a management entity, may exist at a given layer. Entities in adjacent layers interact through the common upper and lower boundaries by passing physical information through service access points (SAPs). A SAP could be compared to a pre-defined 'postbox' where one layer would collect data from the previous layer. The relationship between layers, entities, functions and SAPs are shown in Figure 2.18.

In the OSI model, the entity in the next higher layer is referred to as the N+1 entity and the entity in the next lower layer as N – 1. The services available to the higher layers are the result of the services provided by all the lower layers.

The functions and capabilities expected at each layer are specified in the model. However, the model does not prescribe how this functionality should be implemented. The focus in the model is on the 'interconnection' and on the information that can be passed over this connection. The OSI model does not concern itself with the internal operations of the systems involved.

Figure 2.19 shows the seven layers of the OSI model.

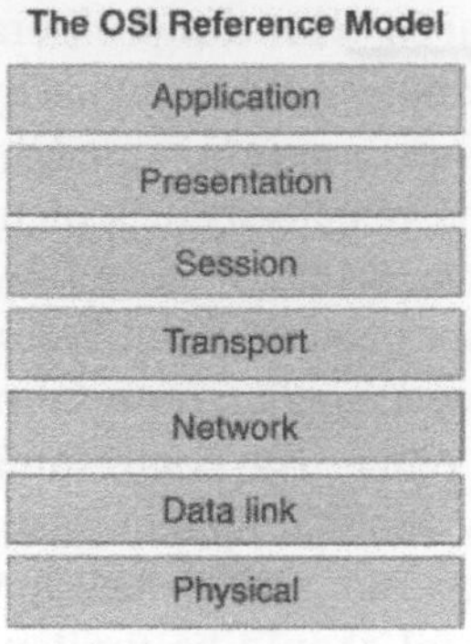

The OSI Reference Model

Application
Presentation
Session
Transport
Network
Data link
Physical

Figure 2.19
The OSI reference model

Typically, each layer on the transmitting side adds header information, or protocol control information (PCI), to the data before passing it on to the next lower layer. In some cases, especially at the lowest level, a trailer may also be added. At each level, this combined data and header 'packet' is termed a protocol data unit or PDU. The headers are used to establish the peer to peer sessions across the sites and some layer implementations use the headers to invoke functions and services at the layers adjacent to the destination layer.

At the receiving site, the opposite occurs with the headers being stripped from the data as it is passed up through the layers. These header and control messages invoke services and a peer to peer logical interaction of entities across the sites. Generally, layers in the same site (i.e. within the same host) communicate in software with parameters passed through primitives, whilst peer layers at different sites communicate with the use of the protocol control information, or headers.

At this stage, it should be quite clear that there is *no* connection or direct communication between the peer layers of the network. Rather, all physical communication is across the physical layer, or the lowest layer of the stack. Communication is down through the protocol stack on the transmitting stack and up through the stack on the receiving stack. Figure 2.20 shows

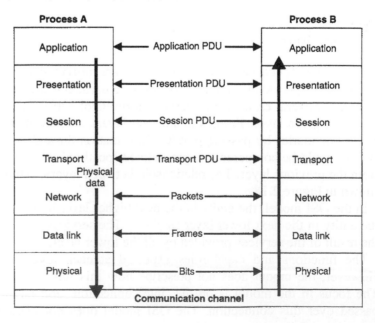

Figure 2.20
Full architecture of the OSI model

the full architecture of the OSI model, whilst Figure 2.21 shows the effects of the addition of PCI to the respective PDUs at each layer. As will be realized, the net effect of this extra information is to reduce the overall bandwidth of the communications channel, since some of the available bandwidth is used to pass control information.

The services provided at each layer of the stack are as follows:

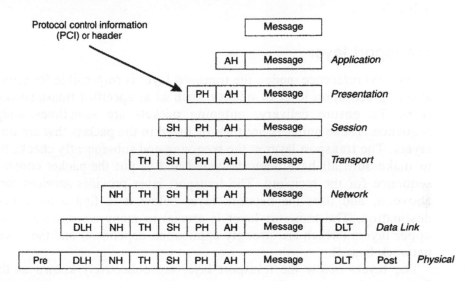

Figure 2.21
OSI message passing

2.13.2   Application layer

The application layer is the topmost layer in the OSI reference model. This layer is responsible for giving applications access to the network. Examples of application-layer tasks include file transfer, electronic mail (e-mail) services, and network management. Application-layer services are much more varied than the services in lower layers, because the entire range of application possibilities is available here. Application programs can get access to the application-layer services in software through application service elements (ASEs). There are a variety of such ASEs; each designed for a class of tasks. To accomplish its tasks, the application layer passes program requests and data to the presentation layer, which is responsible for encoding the application layer's data in the appropriate form.

2.13.3   Presentation layer

The presentation layer is responsible for presenting information in a manner suitable for the applications or users dealing with the information. Functions such as data conversion from EBCDIC to ASCII (or vice versa), use of special graphics or character sets, data compression or expansion, and data encryption or decryption are carried out at this layer. The presentation layer provides services for the application layer above it, and uses the session layer below it. In practice, the presentation layer rarely appears in pure form, and it is the least well defined of the OSI layers. Application- or session-layer programs will often encompass some or all of the presentation layer functions.

2.13.4   Session layer

The session layer is responsible for synchronizing and sequencing the dialogue and packets in a network connection. This layer is also responsible for making sure that the connection is maintained until the transmission is complete, and ensuring that appropriate security measures are taken during a 'session' (that is, a connection).

The session layer is used by the presentation layer above it, and uses the transport layer below it.

2.13.5   Transport layer

In the OSI reference model, the transport layer is responsible for providing data transfer at an agreed-upon level of quality, such as at specified transmission speeds and error rates. To ensure delivery, outgoing packets are sometimes assigned numbers in sequence. These numbers are then included in the packets that are transmitted by lower layers. The transport layer at the receiving end subsequently checks the packet numbers to make sure all have been delivered and to put the packet contents into the proper sequence for the recipient. The transport layer provides services for the session layer above it, and uses the network layer below it to find a route between source and destination. The transport layer is crucial in many ways, because it sits between the upper layers (which are strongly application-dependent) and the lower ones (which are network-based).

The layers below the transport layer are collectively known as the 'subnet' layers. Depending on how well (or not) they perform their function, the transport layer has to interfere less (or more) in order to maintain a reliable connection.

Three types of subnet services (i.e. the service supplied by the underlying physical network between two hosts) are distinguished in the OSI model:

- *Type A*:   Very reliable, connection-oriented service.
- *Type B*:   Unreliable, connection-oriented service.
- *Type C*:   Unreliable, possibly connectionless service.

To provide the capabilities required for the above service types, several classes of transport layer protocols have been defined in the OSI model:

- TP0 (Transfer Protocol Class 0), which is the simplest protocol. It assumes type A service; that is, a subnet that does most of the work for the transport layer. Because the subnet is reliable, TP0 requires neither error detection or error correction. Because the connection is connection-oriented, packets do not need to be numbered before transmission.
- TP1 (Transfer Protocol Class 1), which assumes a type B subnet; that is, one that may be unreliable. To deal with this, TP1 provides its own error detection, along with facilities for getting the sender to retransmit any erroneous packets.
- TP2 (Transfer Protocol Class 2), which also assumes a type A subnet. However, TP2 can multiplex transmissions, so that multiple transport connections can be sustained over the single network connection.
- TP3 (Transfer Protocol Class 3), which also assumes a type B subnet. TP3 can also multiplex transmissions, so that this protocol has the capabilities of TP1 and TP2.
- TP4 (Transfer Protocol Class 4), which is the most powerful protocol, in that it makes minimal assumptions about the capabilities or reliability of the subnet. TP4 is the only one of the OSI transport-layer protocols that supports connectionless service.

2.13.6   Network layer

The network layer is the third lowest layer, or the uppermost subnet layer. It is responsible for the following tasks:

- Determining addresses or translating from hardware to network addresses. These addresses may be on a local network or they may refer to networks located elsewhere on an internetwork. One of the functions of the network layer is, in fact, to provide capabilities needed to communicate on an internetwork.
- Finding a route between a source and a destination node or between two intermediate devices.
- Fragmentation of large packets of data into frames which are small enough to be transmitted by the underlying data link layer (fragmentation). The corresponding network layer at the receiving node undertakes re-assembly of the packet.

2.13.7   Data link layer

The data link layer is responsible for creating, transmitting, and receiving data packets. It provides services for the various protocols at the network layer, and uses the physical layer to transmit or receive material. The data link layer creates packets appropriate for the network architecture being used. Requests and data from the network layer are part of the data in these packets (or frames, as they are often called at this layer). These packets are passed down to the physical layer and from there, the data is transmitted to the physical layer on the destination machine. Network architectures (such as Ethernet, ARCnet, token ring, and FDDI) encompass the data-link and physical layers, which is why these architectures support services at the data-link level. These architectures also represent the most common protocols used at the data-link level.

The IEEE (802.x) networking working groups have refined the data-link layer into two sub-layers: the logical-link control (LLC) sub-layer at the top and the media-access control (MAC) sub-layer at the bottom. The LLC sub-layer must provide an interface for the network layer protocols, and control the logical communication with its peer at the receiving side. The MAC sub-layer must provide access to a particular physical encoding and transport scheme.

2.13.8   Physical layer

The physical layer is the lowest layer in the OSI reference model. This layer gets data packets from the data-link layer above it, and converts the contents of these packets into a series of electrical signals that represent 0 and 1 values in a digital transmission. These signals are sent across a transmission medium to the physical layer at the receiving end. At the destination, the physical layer converts the electrical signals into a series of bit values. These values are grouped into packets and passed up to the data-link layer.

Transmission properties defined

The mechanical and electrical properties of the transmission medium are defined at this level. These include the following:

- *The type of cable and connectors used*:   The cable may be coaxial, twisted-pair, or fiber optic. The types of connectors depend on the type of cable.
- *The pin assignments for the cable and connectors*:   Pin assignments depend on the type of cable and also on the network architecture being used.

- *Format for the electrical signals*:   The encoding scheme used to signal 0 and 1 values in a digital transmission or particular values in an analog transmission depend on the network architecture being used. Most networks use digital signaling, and most use some form of Manchester encoding for the signal.

Note that this layer does *not* include the specifications of the actual medium used, but rather *which* medium should be used, and *how*. The specifications of, for example, Unshielded Twisted Pair as used by Ethernet is contained in specification EIA/TIA 568.

2.13.9   Summary of OSI layers

Here follows, in conclusion, a brief summary of the seven layers.

- *Application*:   The provision of network services to the user's application programs.
  Note: the actual application programs do *not* reside here.
- *Presentation*:   Primarily takes care of data representation (including encryption).
- *Session:*   Control of the communications (sessions) between the users.
- *Transport*:   The management of the communications between the two end systems.
- *Network*:   Primarily responsible for the routing of messages.
- *Data Link*:   Responsible for assembling and sending a frame of data from one system to another.
- *Physical*:   Defines the electrical signals and mechanical connections at the physical level.

3

Transmission media

Objectives

When you have completed studying this chapter you should be able to:

- Describe the effects of attenuation on a digital signal
- Discuss the impact of induced magnetic fields on a twisted-pair cable
- Explain the various methods used to extend a subscriber's loop
- Discuss the implications of using loading coils on a subscriber plant
- Explain 'tilt' and how it is overcome in coaxial cable systems
- Identify the component parts of a fiber-optic cable
- Discuss the difference between single-and multi-mode fibers
- Understand the calculation of link budgets on fiber-optic systems
- Compare the advantages of fiber-optic systems with those of radio
- Explain how radio fade margins impact on system availability
- List the components in a VSAT link.

3.1   Introduction

In this chapter some of the different types of transmission media used for physically conveying signals from one point to another will be examined. The approach taken will be to explain the fundamental method of operation of each of these transmission media types, introduce the various system components and discuss the application for each type. Some of the main bearer design considerations will be discussed to enable the reader to make an informed decision as to which type of media to use for a particular application. The discussion will commence with systems guided over a physical bearer; namely twisted-pair and coaxial copper, fiber-optic cables and the power distribution system. The discussion will then move on to wireless systems; namely microwave radio systems, satellite systems and infrared transmission, which require no specific bearer and radiate their signals as electromagnetic waves. The emphasis in this chapter is on the fundamental transmission bearer systems, Chapter 6 discusses the methods of carrying information, both analog and digital, on these bearer systems.

3.2   Basic cable transmission parameters

Ideally, the signal at the end of a length of cable should be the same as at the beginning. Unfortunately, this is not true in practical cables. Signal quality degrades for several reasons, including attenuation, crosstalk and impedance mismatches. Any transmission also consists of signal and noise components.

Attenuation

All signals degrade when transmitted over a distance through any medium. This is because the amplitude of the signal decreases as the medium resists the flow of energy. Attenuation is the decrease in signal strength, measured in decibels (dB) per unit length. Attenuation in copper cables is primarily dependent on the resistance of the conductors. This causes some of the electrical energy in the signal to be converted to heat energy as the signal progresses along the cable, resulting in a continuous decrease in the amplitude of the signal. In fiber-optic systems the transmitted light is similarly attenuated, with light being absorbed by the natural resistance properties of the glass fiber.

Attenuation increases with frequency. This causes attenuation to any practical signals containing a range of frequencies. Figure 3.1 illustrates a digital signal with very fast rising edges to the pulses needing very high frequency components to reproduce accurately. When such a signal is sent through a transmission channel the higher frequencies attenuate more than the lower frequencies. As a result the transitions in the pulses decrease progressively as the signal travels along the channel. This figure also illustrates the signal attenuation with the amplitude of the signal decreasing with distance.

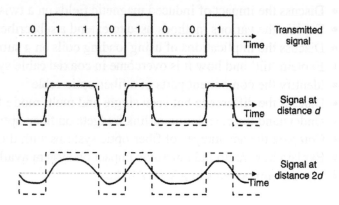

Figure 3.1
Signal attenuation effects

To allow signal attenuation, a limit is set for the maximum length of the communications channel. This is to ensure that the attenuated signal arriving at the receiver is of sufficient amplitude to be reliably detected and correctly interpreted. If the channel is longer than this maximum length, amplifiers or repeaters must be used at regular intervals along the channel to restore the signal to acceptable levels. This is illustrated in Figure 3.2. Since attenuation is sensitive to frequency, some situations require equalizers to boost different-frequency signals to appropriate amount.

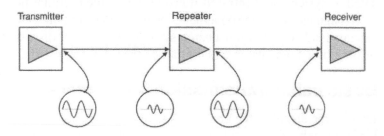

Figure 3.2
Use of signal repeaters

Crosstalk

Crosstalk is interference in the form of a signal from a neighboring cable or circuit; for example, signals on different pairs of twisted wire in a twisted-pair cable may interfere with each other. A commonly used measure of this interference in twisted-pair cable is near-end crosstalk (NEXT), which is represented in decibels. Higher the decibel value, less the crosstalk and better the cable. Additional shielding between the carrier wire and the outside world is the most common way to decrease the effects of crosstalk. Crosstalk is nonexistent for fiber-optic cables.

Characteristic impedance

The impedance of a cable, is defined as the opposition to the flow of electrical energy at a particular frequency. The characteristic impedance of a cable is the value of impedance which characterizes that cable. The characteristic impedance is the input impedance of the cable seen when it is terminated in the characterisic impedance, as shown in Figure 3.3. Such a cable then appears electrically to be infinitely long and has no signal reflected from the termination. If one cable is connected to another of differing characteristic impedance, then signals are reflected at their interface. These reflections cause interference with the data signals and must be avoided by using cables of the same characteristic impedance.

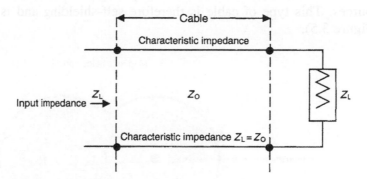

Figure 3.3
Characteristic impedance

Noise performance

As the signals pass through a communication channel the molecules in the transmission medium vibrate and emit random electromagnetic signals as noise. The strength of the transmitted signal is normally large relative to the noise signal. However, as the signal travels through the channel and is attenuated, its level can approach that of the noise. When the level of the wanted signal is not significantly higher than the background noise, the receiver cannot separate the data from the noise and communication errors occur. An important parameter of the channel is the ratio of the power of the received signal (S) to the power of the noise signal (N). The ratio S/N is called the signal-to-noise ratio, which is normally expressed in decibels as follows:

$$\text{Signal-to-noise ratio} = 10 \log_{10}(S/N) \text{ dB}$$

Where

    S = signal power in W

    N = noise power in W.

3.3   Twisted-pair copper cables

Twisted-pair cable is widely used, inexpensive, and easy to install. Twisted-pair cable used for telecommunication purposes is generally unshielded (UTP). It can transmit data at an acceptable rate – up to 100 Mbps in some network architectures. The most common twisted-pair wiring is telephone cable, which is unshielded and is usually voice-grade, rather than the higher-quality category 5 cable used for data networks. In a twisted-pair cable, two conductor wires are wrapped around each other. The information signal is transmitted as a difference voltage between the two conductor wires. The current flows in opposite directions in each wire of the active circuit, as shown in Figure 3.4.

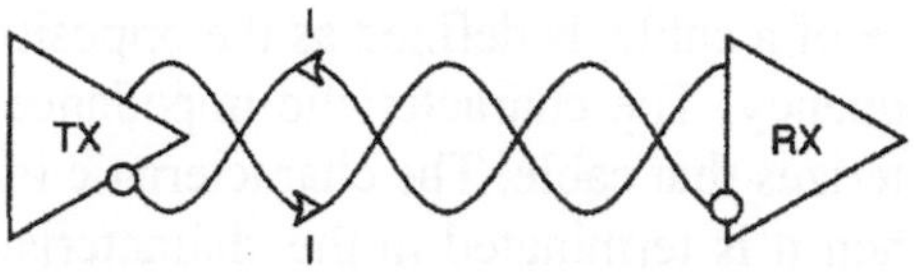

Figure 3.4
Current flow in a twisted-pair cable

Since these currents are equal and opposite and in close proximity, their magnetic fields cancel each other, and also cancel any magnetic interference caused by outside noise sources. This type of cable is therefore self-shielding and is less prone to interference (Figure 3.5).

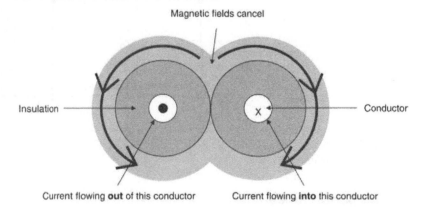

Figure 3.5
Shielding of twisted-pair cables

Twisting the pairs of wires minimizes crosstalk between pairs. The twists also help deal with electromagnetic interference (EMI) and radio frequency interference (RFI), as well as balancing the mutual capacitance of the cable pair. The performance of a twisted-pair cable can be influenced by changing the number of twists per meter in a wire pair. Each of the pairs in a 4-pair category 5 cable, for example has a different twist rate to reduce the crosstalk between them.

3.3.1   Components of twisted-pair cable

A twisted-pair cable is made from pairs as illustrated in Figure 3.7 and the pairs are bundled to form a cable as illustrated in Figure 3.6. The components are:

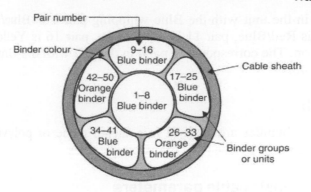

Figure 3.6
Twisted-pair cable components

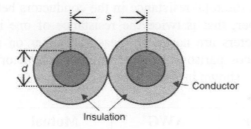

Figure 3.7
Twisted-pair construction

Conductor wires

The signal wires for this cable are conductors that are wrapped around each other. The conductor wires are usually made of copper. They are normally solid (consisting of a single wire) but can be stranded (consisting of many thin wires wrapped tightly together) for short, flexible patch cords. A twisted-pair cable contains multiple twisted-pairs; 2, 4, 7, 15, 25, 50, 100, 200, 300 or 400 pairs are common sizes. Bundles of wire pairs are twisted around each other then whipped with color coded binder tapes to form a cable unit or a binder group. The cable is made out of numbers of binder group's wrapped around each other. The number of pairs in each binder group and the number of groups are dependent on the size of the cable. For example, the 50-pair cable illustrated in Figure 3.6 is made up of six binder groups of 8 or 9 pairs.

Insulation

Each of the individual conductor in the cable is surrounded by an insulating layer. This layer needs sufficient thickness to maintain the insulation at the maximum working voltage of the circuit. For telecommunication cables the normal maximum working voltage is typically 140 V AC, but the cables may be subjected to insulation testing at 500 V. On the other hand the insulation needs to be sufficiently thin to enable the cable to have the required impedance whilst minimizing the capacitance between the conductors. Modern telecommunication cables often use foamed polythene (PE) insulation for this purpose. Small size distribution cables have color coded insulation to readily identify the pair and to ensure the correct polarity of the line. The A-leg is marked White, Red, Black, Yellow or Violet in sequence to identify which group of five pairs, while the B-leg is consecutively coded Blue, Orange, Green, Brown and Slate. Colored whippings identify the 25-pair unit, for example Blue 1–25, Orange 26–50, Green 51–75 and Brown 76–100.

Hence in the unit with the Blue whipping pair 1 is Blue/White, pair 2 is Orange/White, pair 6 is Red/Blue, pair 11 is Black/Blue, pair 16 is Yellow/Blue, pair 21 is Violet/Blue and so on. The corresponding pairs in the unit with the Orange whipping would be 26, 27, 31 and 46.

Sheath

The wire bundles are encased in a sheath made of polyvinylchloride (PVC) for internal cables or polyethylene (PE) for external cables.

3.3.2   Twisted-pair cable parameters

Loop resistance

The resistance of a copper cable pair depends on the diameter of the conductors. Loop resistance is the total resistance in the conductors between the Central Office switch and the subscriber, that is twice the resistance of one individual conductor of that length. Cable diameters are normally measured either on millimeters or American wire gage (AWG). A comparison of these different conductor sizes and their corresponding loop resistances is shown in Table 3.1.

Diameter (mm)	AWG (Approx.)	Mutual Capacitance (nF/km)	Loop Resistance (Ω/km)	Attenuation @1000 Hz (dB/km)
0.32	28	40	433	2.03
		50	433	2.13
0.4	26	40	277	1.62
		50	277	1.69
0.5	24	40	177	1.30
		50	177	1.33
0.63	22	40	110	1.01
		50	110	1.06
0.9	19	40	55	0.72
		50	55	0.75

Table 3.1
Typical twisted-pair cable properties

Characteristic impedance

The characteristic impedance of the twisted-pair cable illustrated in Figure 3.7 is given by the following formula:

$$Z_0 = \frac{276}{\sqrt{k}} \log\left(\frac{2s}{d}\right)$$

Where
   k is the dielectric constant of the insulation
   s is the conductor spacing
   d is the conductor diameter.

Telecommunication cables have characteristic impedance of about 100 Ω.

Mutual capacitance

The mutual capacitance of the cable pair depends on the conductor geometry and the dielectric constant of the insulation. Typical cables range from 40 to 50 nF/km as shown in Table 3.1.

Attenuation

The attenuation of twisted-pair cable varies with frequency as mentioned earlier. Most subscribers' loop calculations are made using attenuation at 1000 Hz, and typical values of attenuation are given in Table 3.1. The variation in attenuation across the speech band is illustrated for two different types of cable in Table 3.2. Twisted-pair copper cables can also be used at frequencies up to 2 MHz as the bearer for high-speed customer access systems such as ADSL as discussed in Chapter 7. Figure 3.8 illustrates the typical high frequency response of a 0.63 mm cable.

Frequency (Hz)	Conductor Diameter (mm)				
	0.32	**0.4**	**0.5**	**0.63**	**0.9**
300	1.17	0.93	0.74	0.59	0.42
600	1.65	1.31	1.04	0.83	0.59
1000	2.13	1.69	1.33	1.06	0.75
1600	2.68	2.12	1.68	1.33	0.92
2000	3.00	2.37	1.87	1.48	1.02
2500	3.34	2.64	2.07	1.64	1.12
3000	3.65	2.88	2.26	1.78	1.21
3400	3.88	3.06	2.40	1.88	1.27

Table 3.2
Typical twisted-pair cable attenuation

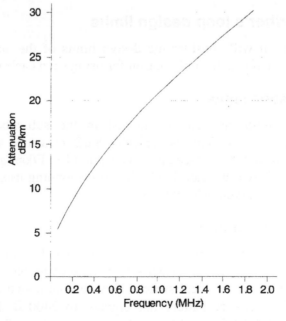

Figure 3.8
Typical high frequency response of a 0.63 mm cable

The attenuation of twisted-pair cables increases with higher frequencies because of skin effect and radiation effects. Skin effect occurs at high bit rates because the current flowing in the wires tends to flow only on the outer surface of the wire thereby reducing its effective cross section. This increases its electrical resistance at high frequencies and results in higher attenuation. At bit rates greater than 1 Mbps significant energy is radiated from the twisted-pair cable resulting in additional signal loss and increased interference to adjacent circuits.

3.3.3   EIA/TIA 568 cable categories

To distinguish varieties of UTP, the Electronic Industries Association/Telecommunications Industries Association (EIA/TIA) has formulated five categories. The electrical specifications for these cables are detailed in the following: EIA/TIA 568A, TSB-36, TSB-40 documents and their successor document SP2840. These categories are:

- *Category 1*:   Voice-grade UTP telephone cable. This describes the cable that has been used for years for telephone communications. Officially, such cable is not considered suitable for data-grade transmissions. In practice, however, it works fine over short distances and under ordinary working conditions.
- *Category 2*:   Voice-grade UTP, although capable of supporting transmission rates of up to 4 Mbps. IBM Type-3 cable falls into this category.
- *Category 3*: Data-grade UTP, used extensively for supporting data transmission rates of up to 10 Mbps. An Ethernet network cabled with twisted-pair requires at least this category of cable.
- *Category 4*:   Data-grade UTP, capable of supporting transmission rates of up to 16 Mbps. An IBM token ring network transmitting at 16 Mbps requires this type of cable.
- *Category 5*:   Data-grade UTP, capable of supporting transmission rates of up to 155 Mbps (but officially only up to 100 Mbps). The 100Base/TX and Gigabit (1000 Mbps) network architectures require such cable.

3.3.4   Subscriber's loop design limits

This section will consider the design limits of the subscriber's loop using twisted-pair cable, since this is the only reason for using such cable in a telecommunication network.

Attenuation limits

The maximum attenuation allowed in the subscriber's loop is set in the national transmission plan. This is typically 8 dB measured at 1000 Hz. Using the values of attenuation for the low-capacitance cable (40 nF/km) given in Table 3.1 for the different conductor sizes, we can derive the corresponding maximum cable distances to meet the 8 dB limit as given in Table 3.3.

Resistance limits

The loop resistance limit or signaling limit is based on the requirement for a minimum current flow in the subscriber's loop to activate both the telephone instrument and the loop detector at the local switch. This minimum loop current is generally 20 mA. Using a -48 V exchange battery this equates to 2400 Ω. From this we need to deduct the resistance of the feed bridge in the exchange, typically 400 Ω, and the resistance of the subscriber's wiring and telephone instrument for which an allowance of 300 Ω is usually made. This leaves the loop resistance limit of 1700 Ω. Using the values of resistance given in Table 3.1 for the different conductor sizes, we can derive the corresponding maximum cable distances to meet the 1700 Ω limit as given in Table 3.4.

Cable Diameter (mm)	AWG (Approx.)	Attenuation @1000 Hz (dB/km)	8-dB Cable Limit (km)
0.32	28	2.03	3.94
0.4	26	1.62	4.94
0.5	24	1.30	6.15
0.63	22	1.01	7.92
0.9	19	0.72	11.11

Table 3.3
8 dB Subscriber loop limits

Cable Diameter (mm)	AWG (Approx.)	Loop Resistance (Ω/km)	1700 Ω Cable Limit (km)
0.32	28	433	3.92
0.4	26	277	6.14
0.5	24	177	9.60
0.63	22	110	15.45
0.9	19	55	30.90

Table 3.4
1700 Ω Subscriber loop limits

It can be seen, by comparing the distances in Tables 3.3 and 3.4 that in general the transmission limit is reached before the signaling limit. Using the worst-case transmission limit for the system design will therefore ensure compliance with the signaling limit. Exceeding the transmission limit is less critical since it would only result in weaker speech, whereas if the signaling limit were exceeded the switch could not detect the subscriber's loop and no call would be possible.

3.3.5 Extending cable distance

The length of the subscriber's loop, as shown above, can be extended over the twisted-pair cable by using one of the following five different methods:

1. *Increasing the diameter of the conductors*: This allows the greater distances shown in Section 3.3.4. This may not be the most cost-effective option as larger diameter copper is more expensive.
2. *The use of voice frequency amplifiers and/or loop extenders*: Voice frequency amplifiers may be inserted to compensate transmission losses in the loop. These require special amplifiers providing gains of up to 7 dB whilst maintaining the DC signaling conditions of the line. When the transmission loss is greatly extended the signaling loss may then become significant. Under these circumstances a loop extender can be used. These provide additional voltage to the feed bridge at the exchange, typically −84 or −96 V DC.
3. *Using inductive loading*: Inductive loading of a subscriber's loop involves insertion of series inductors, called loading coils, at regular intervals along the

cable. These reduce the voice frequency losses. The loading coils are used in various configurations, and typically have inductance of either 44 or 88 mH (milliHenries) with regular spacing of about 1370 or 1830 m. The disadvantage in loading of cable pairs is that their high frequency response is seriously affected and the pairs cannot be used for any digital transmission systems without removal of the loading coils. The use of loading coils is illustrated in Figure 3.9.

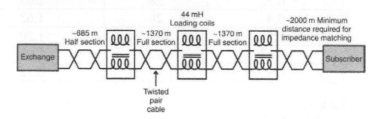

Figure 3.9
Example of the use of loading coils

4. *Using digital transmission systems*:   Digital transmission systems are available for installation on copper cables. These normally use pulse code modulation (PCM) and provide 24 circuits (DS1)(T1) at 1.544 Mbps or 30 circuits (E1) at 2.048 Mbps using regenerators at intervals of about 1800 m which operate over two cable pairs. The cable pairs are used to power feed the regenerators as well as provide separate transmit and receive circuits. The transmit and receive pairs usually need to be located in the different binder groups in the cable, or separate cables, to minimize crosstalk.

The digital transmission system can be used to connect directly to subscriber's distribution cable. These systems are called subscribers PCM and incorporate the subscriber's line circuits functions in the remote terminal. The losses in the digital path are typically 2 dB regardless of the length of the circuit, enabling very remote subscribers to be served by this method. A subscribers PCM system is illustrated in Figure 3.10.

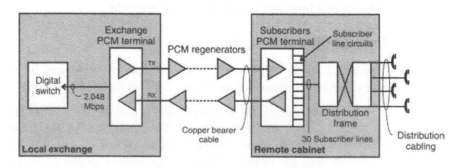

Figure 3.10
Subscribers PCM system

5. *Using remote concentrators or remote switches with digital feeder circuits*: Alternatively the digital transmission system can be connected to a remote concentrator or switch. Here the traffic from all the subscribers at the remote terminal is concentrated on to the digital transmission system. The remote terminal may simply be a concentrator in which two subscribers off the terminal are

connected together at the host local exchange. Alternatively the remote terminal may incorporate local switching, whereby they can be connected together without using two feeder circuits back to the host local exchange. These alternatives are illustrated in Figure 3.11.

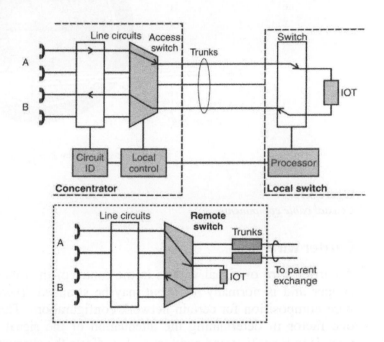

Figure 3.11
Remote concentrators and remote switches

3.3.6   Applications for twisted-pair cable

Twisted-pair cable is almost exclusively used for connection of the subscriber to the local switch. A major part of the investment in the telephone network is in the provision of the subscriber loop plant. Clearly the most cost-effective option needs to be taken in this area. By utilizing digital feeder systems using either PCM over the existing copper, or for new work fiber-optic cables, it is possible to utilize smaller diameter copper distribution cables. Consideration should also be given to the provision of high-speed data circuits for residential customers, particularly for Internet access.

3.4   Coaxial cable

Coaxial cable, often called coax, is used for radio frequency and data transmission. The cable is remarkably stable in terms of its electrical properties at frequencies below 4 GHz (gigahertz), and this makes the cable popular as the connection between a radio and its antenna. Before fiber-optic cables were available, telephone companies used coaxial cable for long distance transmission systems. These have now almost all been replaced with fiber-optic systems. The only significant use of coaxial cables today is for cable television applications, particularly as hybrid fiber coax (HFC) systems discussed in Chapter 7.

3.4.1   Coaxial cable components

A coaxial cable consists of the following layers (moving outward from the center) as shown in Figure 3.12.

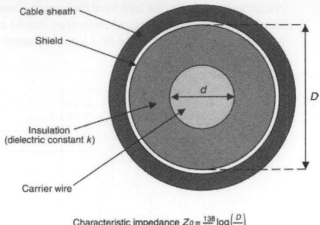

Characteristic impedance $Zo = \frac{138}{\sqrt{K}} \log\left(\frac{D}{d}\right)$

Figure 3.12
Coaxial cable components

Carrier wire

A carrier wire or signal wire is in the center of the cable. This wire is usually made of copper and is normally solid but may be stranded. There are restrictions regarding the wire composition for certain network configurations. The diameter of the signal wire is one factor in determining the attenuation of the signal over distance. The number of strands in a multi-strand conductor also affects the attenuation.

Insulation

An insulation layer consists of a dielectric around the carrier wire. The dielectric is often air, with a helical spacer made of polyethylene or Teflon. For short distance cables foamed or solid dielectric is used.

Foil shield

A thin foil shield around the dielectric. This shield usually consists of aluminum bonded to both sides of a mylar tape. Not all coaxial cables have foil shielding. Some have two foil shield layers, interspersed with copper braid shield layers.

Braid shield

A braid, or mesh, conductor, made of copper or aluminum, surrounds the insulation and foil shield. This conductor is normally connected to ground to create an electrostatic shield for the carrier wire. Together with any foil shield, the earthed braid protects the carrier wire from electromagnetic interference (EMI) and radio frequency interference (RFI). You should carefully note that the braid and foil shields provide good protection against electrostatic interference when earthed correctly, but little protection against magnetic interference.

Sheath

The outer cover of the cable provides physical protection for the cable. This is usually made of polyethylene (PE) for outdoor applications and polyvinyl chloride (PVC) for internal cabling.

The shield layers surrounding the carrier wire also help prevent signal loss due to radiation from the carrier wire. The signal and shield wires are concentric, or coaxial and hence the name.

3.4.2 Coaxial cable parameters

Attenuation

Attenuation of coaxial cables varies with frequency. A typical response curve for a coaxial cable used for cable TV applications is shown in Figure 3.13. It can be seen that there is an exponential increase in attenuation as the frequency increases. In the CATV industry this parameter is referred to as 'tilt'. Practical transmission systems need a relatively flat frequency response across the working bandwidth. To achieve this the system amplifiers provide equalization by amplifying the lower frequencies less than the higher frequencies.

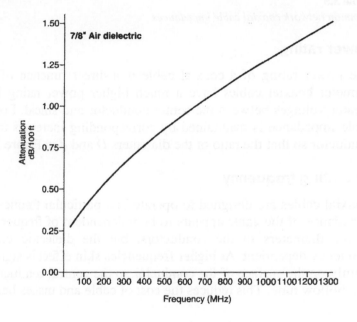

Figure 3.13
Frequency response curve for CATV coaxial cable

Impedance

Impedance is a measure of opposition to the flow of alternating current. The properties of the dielectric between the carrier wire and the braid help to determine the cable's impedance. The impedance of the coaxial cable in Figure 3.12 is given by the formula:

$$Z_0 = \frac{138}{\sqrt{k}} \log\left(\frac{D}{d}\right)$$

Where k is the dielectric constant of the insulation.

Impedance matching is vital in all coaxial cable systems. The transmitting device needs its impedance matched to that of the coaxial cable to ensure maximum power transfer. If the impedances are not matched then some of the applied power would be reflected and the balance transferred. Maximum power transfer is obtained when the impedances of the transmitter and cable or cable and load are matched. A terminating resistance or load is always required at the end of any coaxial cable to match the characteristic impedance of the cable and prevent reflections. Special connector hardware is therefore required to join two coaxial cables together to minimize any power loss at the joint.

Most coaxial cables have a RG (Recommended Gage) rating and cable with the same RG rating from different manufacturers can be safely mixed. In networks, the characteristic cable impedances range from 50 Ω (for an Ethernet architecture) to 93 Ω (for an ARCnet architecture (Table 3.5)).

Recommended Gage	Application	Characteristic Impedance (Ω)
RG-8	10Base-5	50 Ω
RG-58	10Base-2	50 Ω
RG-59	CATV	75 Ω
RG-62	ARCnet	93 Ω

Table 3.5
Common network coaxial cable impedances

Power rating

The power rating of a coaxial cable is a direct function of the cable diameter. Larger diameter coaxial cables have a much higher power rating because they can withstand greater voltages between the center conductor and shield. For larger diameter cables, the cable impedance is maintained by corresponding increases in the diameter of the central conductor so that the ratio of the diameters D and d in Figure 3.12 is maintained.

Operating frequency

Coaxial cables are designed to operate in a particular frequency band. At first glance the impedance of the cable appears to be independent of frequency, being based on the ratio of the diameters of the conductors, but the dielectric constant of the insulation is frequency dependent. At higher frequencies skin effect is significant and the energy flows mainly on the surface of the conductor so the central conductor is often made in the form of a hollow tube. This reduces the cost of cable and makes bending easier.

3.4.3   Coaxial cable designations

Listed below are some of the available coaxial cable types.

RG-6

Used as a drop cable for CATV transmissions. It is a broadband cable, and is often quad-shielded, has 75 Ω impedance.

RG-8

Used for thick Ethernet. It has 50 Ω impedance. It is also known as N-Series Ethernet cable.

RG-11

Used for the main CATV trunk. It has 75 Ω impedance and is a broadband cable. This cable is often quad-shielded (with foil/braid/foil/braid around the signal wire and dielectric) to protect the signal wire even under the worst operating conditions.

RG-58

Used for thin Ethernet. It has 50 Ω impedance and uses a BNC connector.

RG-59

Used for ARCnet. It has 75 Ω impedance and uses BNC connectors. This type of cable is used for broadband connections and also by cable companies to connect the cable network to an individual household.

RG-62

Used for ARCnet. It has 93 Ω impedance and uses BNC connectors. This cable is also used to connect terminals to terminal controllers in IBM's 3270 systems.

3.4.4   Broadband transmission on coaxial cable

Most signals transmitted on coaxial cable are broadband FDM. In these systems, the system bandwidth is divided into a number of non-overlapping frequency slots each of which carries a single channel. Each of the channels operate at a different RF carrier frequency and the system multiplexer modulates each of these with one system RF carrier. In this way the individual channels are effectively stacked up across the system bandwidth, as discussed in Chapter 2. For CATV systems the bandwidth extends from 55–270 MHz for 30 channels through to 55–550 MHz for 78 channels. Consider a cable TV distribution system as an example. Here the broadband FDM signal is transmitted from the Headend (studio) over the coaxial cable and the signal is boosted by trunk amplifiers at regular intervals along the cable. Each amplifier equalizes the signal and boosts its amplitude before sending the signal on to the next amplifier. However each amplifier contributes some noise to the signal which accumulates as the signal traverses many amplifiers. This limits the number of amplifiers that can be used between the Headend and the subscriber. Signals are tapped off the trunk system by bridge amplifiers then distributed to each of the subscribers by means of taps. At the subscriber's premises the signals are presented to a set-top cable box containing the demultiplexer, which demodulates the appropriate channel and presents the original channel information to the TV set. This is illustrated in Figure 3.14.

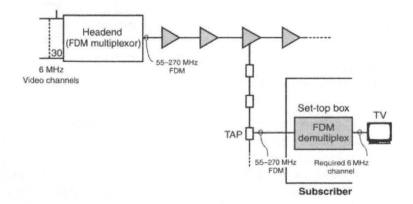

Figure 3.14
CATV distribution system schematic

Applications for coaxial cable

Today coaxial cable is almost exclusively used for cable television systems. For telecommunication applications the hybrid fiber coax systems (HFC) as discussed in Chapter 7 enable two-way high speed data to be sent in addition to the cable television distribution. This makes these systems ideal for providing high-speed Internet access to residential cable customers. In the long term these systems are likely to be replaced by fiber to the home (FTTH) or by fiber to the curb (FTTC) systems.

3.5   Fiber-optic cable

Fiber-optic communication uses light signals and so transmissions are not subject to electromagnetic interference. Fiber-optic cables act as a waveguide for light, with all the energy guided through the central core of the cable. The light is guided due to the presence of a lower refractive index cladding surrounding the central core. None of the energy in the signal is able to escape into the cladding and no energy is able to enter the core from any external sources. The composition of the cable is shown in Figure 3.15.

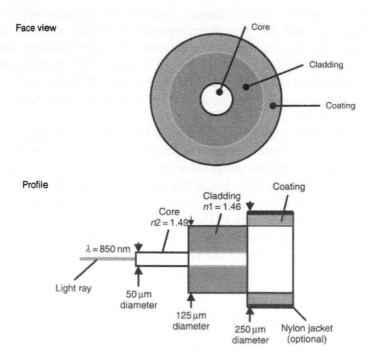

Figure 3.15
Components of an optical fiber

Since little of the light signal is absorbed in the glass core fiber-optic cables can be used for much longer distances before the signal must be amplified, or repeated. Some fiber-optic segments can be many kilometers long before a repeater is needed. Data transmission using a fiber-optic cable is many times faster than electrical methods and speeds of over 10 Gbps are possible. Fiber-optic cables deliver more reliable transmissions over greater distances, although at a somewhat greater cost. Cables of this type differ in their physical dimensions and composition and in the wavelength(s) of light which the system transmits.

3.5.1   Fiber-optic cable components

The major components of a fiber-optic cable are core, cladding, buffer, strength members and jacket. Some types of fiber-optic cable even include a conductive copper wire that can be used to provide power to a repeater.

Fiber core

The core of an optical fiber carries the light signal. The most common core sizes are 50 and 62.5 μm (microns) which are used in multimode cables. Fibers of 8.5 micron size are used in single mode systems.

Cladding

The cladding is a larger of glass surrounding the core. The core and cladding are actually manufactured as a single unit. The cladding is a protective layer with a lower index of refraction than the core. The lower index means the light that hits the core walls will be redirected back to continue on its path. The cladding diameter is typically 125 microns.

Fiber-optic coating

The coating of an optical fiber is made of one or more layers of plastic surrounding the cladding. The coating helps strengthen the cable, thereby decreasing the likelihood of microcracks, which can eventually break the fiber. The coating also protects the core and cladding from potential invasion by water or other materials in the operating environment. The coating typically doubles the diameter of the fiber.

Buffer

A buffer can be loose or tight. A loose buffer is a rigid tube of plastic with one or more fibers (consisting of core and cladding) loosely running through it. The tube takes all the stresses applied to the cable, isolating the fiber from these stresses. A tight buffer is a layer of tough plastic, fitting snugly around the fiber. A tight buffer can protect the fibers from stress due to pressure and impact, but not from changes in temperature.

Strength members

Fiber-optic cable also has strength members, which are strands of very tough material (such as steel, fiberglass, or Kevlar) that provide tensile strength for the cable. Each of these materials has advantages and drawbacks. For example, steel conducts electricity making the cable vulnerable to lightning, which will not disrupt an optical signal but may seriously damage the cable or equipment.

Cable sheath

The sheath of a fiber-optic cable is an outer casing that provides primary mechanical protection, as with electrical cable.

3.5.2   Fiber-optic cable parameters

Attenuation

The attenuation of a multimode fiber depends on the wavelength and the fiber construction, and ranges from around 3–8 dB/km at 850 nm and 1–3 dB/km at 1300 nm.

The attenuation of single-mode fiber ranges from around 0.4–0.6 dB/km at 1300 nm and 0.25–0.35 dB/km at 1550 nm.

Diameter

The fiber diameter is either 50 or 62.5 microns for multimode fiber or 8.5 microns for single mode.

- *Multimode fibers (50 or 62.5 microns)*:   In multimode fibers a beam of light has room to follow multiple paths, which are called modes, through the core. Multiple modes in a transmission produce signal distortion at the receiving end, due to the difference in arrival time between the fastest and slowest of the alternate light paths.
- *Single-mode fibers (8.5 microns)*:   In a single-mode fiber, the core is so narrow that the light can take only a single path through it. Single-mode fiber has the least signal attenuation, usually less than 0.5 dB/km. This type of cable is the most difficult to install, because it requires precise alignment of the system components and the light sources and detectors are very expensive. However, transmission speeds of 50 Gbps and higher are possible.

Wavelength

Fiber-optic systems today operate in one of three wavelength bands; 850, 1300 or 1550 nm. The shorter wavelengths have a greater attenuation than the longer wavelengths. Short-haul systems tend to use the 850 or 1300 nm wavelengths with multimode cable and light emitting diode (LED) light sources. The 1550 nm fibers are used almost exclusively with the long-distance systems using single-mode fiber and laser light sources.

Bandwidth

The bandwidth of a fiber is given as the range of frequencies across which the output power is maintained within 3 dB of the nominal output. It is quoted as the product of the frequencies of bandwidth multiplied by distance, for example 500 MHz-km. This means that 500 MHz of bandwidth is available over a distance of 1 km or 100 MHz of bandwidth over 5 km.

Dispersion

Modal dispersion is measured as nanoseconds of pulse spread per kilometer (ns/km). The value also imposes an upper limit on the bandwidth, since the duration of a signal must be larger than the nanoseconds of a tail value. With step-index fiber, expect between 15 and 30 ns/km. Note that a modal dispersion of 20 ns/km yields a bandwidth of less than 50 Mbps. There is no modal dispersion in single-mode fibers, because only one mode is involved.

Chromatic dispersion occurs in single mode cables and is measured as the spread of the pulses in picoseconds for each nanometer of spectral spread of the pulse and for each kilometer traveled. This is the only dispersion effect in single mode cables and typical values are in the order of 3.5 ps/nm-km at 1300 nm and 20 ps/nm-km at 1550 nm.

3.5.3   Types of fiber-optic cable

One reason optical fiber makes such a good transmission medium is because the different indexes of refraction for the cladding and core help to contain the light signal within the core, producing a waveguide for the light. Fiber can be constructed by changing abruptly from the core refractive index to that of the cladding, or this change can be made gradually (Figure 3.16). The two main types of multimode fiber differ in this respect.

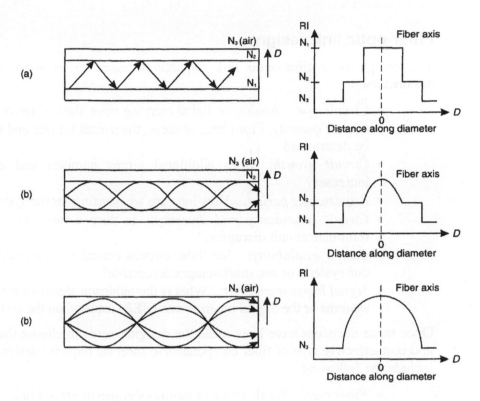

Figure 3.16
Fiber refractive index profiles

Step-index cable

Cable with an abrupt change in refraction index is called step-index cable. In step-index cable, the change is made in a single step. Single-step multimode cable uses this method, and it is the simplest, least expensive type of fiber-optic cable. It is also the easiest to install. The core is usually 50 or 62.5 microns in diameter; the cladding is normally 125 microns. The core width gives light quite a bit of room to bounce around in, and the attenuation is high (at least for fiber-optic cable): between 10 and 50 dB/km. Transmission speeds up to 10 Mbps over 1 km are possible.

Graded-index cable

Cable with a gradual change in refraction index is called graded-index cable, or graded-index multimode. This fiber-optic cable type has a relatively wide core, like single-step multimode cable. The change occurs gradually and involves several layers, each with a slightly lower index of refraction. A gradation of refraction indexes controls the light signal better than the step-index method. As a result, the attenuation is lower, usually less than 15 dB/km. Similarly, the modal dispersion can be 1 ns/km and lower, which allows more than ten times the bandwidth of step-index cable (Figure 3.16).

Fiber designations

Optical fibers are specified in terms of their core, cladding and coating diameters. For example, a 62.5/125/250 fiber has a core with a 62.5 micron diameter with a cladding of 125 microns and a coating of 250 microns.

3.5.4 Fiber-optic link design

The design of a fiber-optic link requires decisions to be made on the following parameters:

- *Circuit type*: Analog or digital carrying voice, data, video or CATV circuits?
- *Circuit quantity*: From both of these, the overall bit rate and system format can be determined.
- *Circuit growth*: Does additional circuit numbers and expected bit rate increases?
- *Provisioning period*: How long are we planning for the system to last?
- *Circuit redundancy and diversity*: Is there a need for diverse routing to minimize circuit disruption?
- *Circuit availability*: Are these circuits critical for the continued operation of our systems or are short outages acceptable?
- *Signal impairment limit*: What is the minimum signal-to-noise ratio for analog systems or the digital bit error rate (BER) required on the system?

Once these decisions have been made the designer needs to choose the most appropriate and cost-effective type of fiber components to meet the required design objectives. These include the following:

- *Fiber type*: Single mode or multimode, step or graded index?
- *Cable type*: How many fibers, tight or loose fiber buffering, cable sheath, armoring, etc.?
- *System wavelength*: 850 nm, 1300 nm, or 1550 nm?
- *Type of optical source*: LED or laser?
- *Type of optical detector*: PIN diode or avalanche photo diode (APD)?
- *Optical amplifiers*: Are any of these required? If so what gain is needed?

Note that alternative methods of circuit provision such as radio systems should also be considered in determining the most appropriate solution.

Once the optical equipment has been decided we need to consult the manufacturer's handbooks to determine:

- *The optical output power of the source*: The parameter required is how much light is coupled into a fiber of the type chosen. This will be quoted in dB m as the end-of-life power.
- *The sensitivity of the chosen optical detector*: This is also known as the receiver threshold and is also quoted in dB m. For the system to operate correctly, at the chosen bit error rate, a signal of this level must arrive at the detector.
- *The dynamic range of the chosen optical detector*: This gives the maximum signal at which the system will operate correctly. If too much light arrives at the optical detector it will be overloaded and the error rate will increase. The dynamic range is quoted in dB and the maximum receiver threshold is calculated by adding this to the minimum receiver threshold.

We are now in a position to calculate the link power budget. We need to take into account in our calculations a reserve of power, typically 6 dB, called the link margin. This allows the following contingencies: variability in the attenuation of the cable, the extra insertion loss of any future splices required to repair the cable and any degradation in the performance of the system components throughout the life of the system.

System gain calculations

The system gain, or power budget, is simply the difference between the power from the optical source coupled into the fiber and the optical receiver sensitivity threshold. From this we need to subtract the losses of all of the components between the transmitter and receiver. These include the losses of the fiber, insertion loss of any connectors, splices, patch cords and any other devices such as a couplers, etc. Where amplifiers are used their gain is added to the power budget. As long as the total losses plus the link margin do not exceed the power budget the design would be satisfactory. A greater link margin would ensure greater system reliability but would incur additional costs. The design of a fiber-optic link is illustrated in Figure 3.17. Check that the received signal level is within the dynamic range of the optical detector. On very short cables it may be necessary to add an optical attenuator to avoid overloading.

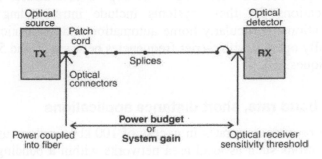

Figure 3.17
Design of a fiber-optic link

3.5.5 Applications for fiber-optic cables

Fiber-optic cables offer the following advantages over other types of transmission media:

- Light signals are impervious to interference from EMI or electrical crosstalk. Light signals do not interfere with other signals. As a result, fiber-optic connections can be used in extremely adverse environments, such as in lift shafts or assembly plants, where powerful motors produce lots of electrical noise.
- Optical fibers have a much wider, flat bandwidth than coaxial cables and equalization of the signals is not required.
- The fiber has a much lower attenuation, so signals can be transmitted much further than with coaxial or twisted-pair cable before amplification is necessary.
- Optical fiber cables do not conduct electricity and so eliminates problems of ground loops, lightning damage and electrical shock when cabling in high-voltage areas.
- Fiber-optic cable is generally much thinner and lighter than electrical cable.
- Fiber-optic cables have greater data security than copper cables.
- Licensing is not required, although a right-of-way for laying the cable is needed.

3.6   Power system carrier

Power line carrier (PLC) utilizes the existing power distribution cables as the transmission medium. This is a very inhospitable environment for data communications because of the presence of frequency dependent attenuation which can rapidly change by up to 60 dB and also considerable interference in the form of impulse noise generated by various electrical devices such as light dimmers, ballasts, motors, etc.

Power line carrier is typically used in three main application areas:

Low baud rate/long-distance applications

These have been traditionally used by the power companies for the load control of customer devices, such as electrical hot water heating. Similar devices have been used on the high voltage power distribution Grid's operated by power utilities. These are able to operate at several hundred baud over distances of hundreds of kilometers using high-level tones superimposed on the 25 kV transmission lines.

Medium baud rate, medium distance applications

These provide data rates up to 50 kbps over distances up to several hundred meters. Applications for these systems include intra-building communication and control applications, particularly home automation and automatic meter reading. These systems typically operate with carrier frequencies between 50 and 535 kHz, using spread spectrum techniques.

High baud rate, short distance applications

These provide data rates in excess of 100 kbps and are used typically for data intensive applications such as local area networks within a building. These typically operate with carrier frequencies between 1.7 and 30 MHz, using spread spectrum techniques.

The CENELEC regulations control PLC operation to minimize interference as follows:

- Data band 95–150 kHz for private users
- 125–140 kHz reserved for devices using unique addressing
- PLC transmitters limited to 500 mW to minimize crosstalk and RF interference.

3.6.1   PLC applications

Interesting developments are being made by some power companies modifying their main power electrical distribution network to enable the transmission of data over the existing cables. This has exciting implications for the provision of high-speed Internet access for the power company customers as well as providing a communications system for remote metering. While power line carrier has been utilized as a point-to-point links on open wire power transmission lines for many years, the provision of similar services on the subscriber's distribution cabling is innovative. Low speed data has also been used on the power distribution systems for use with ripple control of electrical hot water heating. The principle of the system is illustrated in Figure 3.18.

The challenge in operating high-speed data access over the power cables is similar to those already discussed for twisted-pair cable, namely attenuation of the signals and interference from other systems. Unfortunately the transformers used in the power distribution system operate as low pass filters, so high frequency data transmissions superimposed on the mains voltage cannot be passed directly. However this can be overcome by the use of high pass data filters bypassing the distribution transformers. The data can be extracted at appropriate access points such as sub-stations then fed to the destination, for example an Internet service provider (ISP), via conventional data circuits particularly fiber-optic cables.

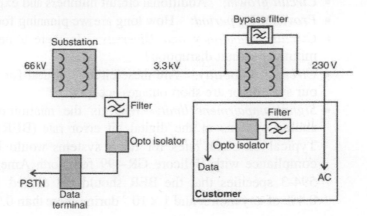

Figure 3.18
Data access over mains distribution wiring

3.6.2 PLC approaches

Many different approaches are used to reliably transmit data in this inhospitable environment. Higher speed systems operate with various forms of spread spectrum and need forward error correction to achieve reliable transmissions. Low speed systems often use some form of synchronization to the zero crossing on the AC power line to take advantage of the lack of interference at this time as well as using the zero crossings as a system clock. The X10 protocol used for home automation systems operates in this way and represents a binary 1 by a 1 ms burst of 120 kHz at the zero crossing point, and a binary 0 by the absence of 120 kHz. These one millisecond bursts need to be transmitted three times to coincide with the zero crossing point of all three phases in a three phase distribution system.

3.7 Microwave radio

In this section the use of line-of-sight (LOS) microwave radio systems as the telecommunication bearers will be discussed.

3.7.1 Radio system design procedure

The design of a line-of-sight microwave link involves four basic processes as follows:

1. Defining the system performance requirements
2. Identification of appropriate transmitting and receiving sites and preparing a radio path profile between the sites
3. Determination of the appropriate antenna heights at each of the sites
4. Selection of the appropriate system components and calculation of the link budget to confirm system operation.

The system performance requirements involves the same parameters already discussed for fiber-optic systems:

- *Circuit type*:   Analog or digital carrying voice, data, video or CATV circuits?
- *Circuit quantity*:   From both of these, the overall bit rate and system format can be determined.
- *Circuit growth*:   Additional circuit numbers and expected bit rate increases?
- *Provisioning period*:   How long are we planning for the system to last?
- *Circuit redundancy and Diversity*:   Is there a need for diverse routing to minimize circuit disruption?
- *Circuit availability*:   Are these circuits critical for the continued operation of our systems or are short outages acceptable?
- *Signal impairment limit*:   What is the minimum signal-to-noise ratio for analog systems or the digital bit error rate (BER) required on the system? Typical bit error rates for radio systems would be less than 2×10^{-10} for compliance with Bellcore GR-499 for North American systems. CCIR Rec. 594-3 specifies that the BER should not exceed 1×10^{-6} during more than 0.4% of any month and 1×10^{-3} during more than 0.054% of any month.

Once these decisions have been made the designer needs to choose the most appropriate and cost-effective type of radio link configuration and specify the components to meet the required design objectives. Alternative methods of circuit provision should also be considered such as fiber-optic systems.

3.7.2   Radio path profile

The next step involves identification of the appropriate transmitting and receiving sites. Terminal sites need to be located close to the equipment they will serve, whereas intermediate repeaters are not constrained. The main site selection criteria will be:

- Availability of the land
- Proximity of mains power
- Proximity of road access and above all
- Feasibility of the radio path.

Preliminary site selection is often done from examination of topographical maps of the area, having a scale of at least 1:50 000. The next step is to draw a straight line along the path between the chosen sites, and read from the topographical map the heights of the two sites, and the respective heights and distances of any potential obstructions along the path. The path profile can now be plotted. This can be done using any suitable computer program, such as 'Path Loss II' or the profile can be prepared manually. In both methods the heights measured from the topographical map, need to be adjusted. Where the obstruction to the radio path is trees then the tree height (often 12 m or 40 ft) needs to be added to the ground level read from the topographical map, together with an allowance for tree growth during the system lifetime (typically 3.0 m or 10 ft).

The topographical map heights make no allowance for the curvature of the earth, so an adjustment is necessary. In addition, the radio path seldom follows a straight line between the transmitter and receiver, but is bent due to refraction in the atmosphere.

Under certain atmospheric conditions the radio path is bent toward the earth whilst under other conditions it is bent away from the earth. This is illustrated in Figure 3.19. This variability in the atmosphere is modeled using the K factor, which is used to modify the earth's radius so as to straighten the radio path. When $K < 1$ the radio beam bends towards the earth and when $K > 1$ the radio beam bends away from the earth. The K factor typically varies between 0.4 and 1.33 as shown in Table 3.6.

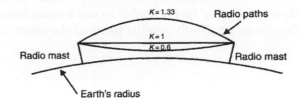

Figure 3.19
Effect of K factor on radio paths

K Factor				
0.4–0.5	**0.5–0.66**	**0.66–1.0**	**1.0–1.33**	**1.33**
Coastal tropical, water	Coastal	Flat, temperature	Mountainous dry	Temperate standard atmosphere
Fog moisture over water	Fog surface layers	Light fog	No fog No surface layers	No fog No surface layers

Table 3.6
K factor

The heights of the intervening obstructions on the radio path need to be modified to compensate for both the ordinary earth curvature and the atmospheric variations by using the K factor in the following formula:

$$h = 0.078 \frac{d_1 d_2}{K} \, \text{m}$$

Where
h = earth curvature adjustment, added to obstacle height
d_1 = distance of point to one end of path in km
d_2 = distance of point to other end of path in km
K = earth curvature factor.

The optimum path clearance above the obstructions is defined to be 0.6 of the Fresnel zone radius (F_1). If this path clearance ($0.6F_1$) is maintained the system behaves as though it were in free space, however if there is less clearance than this, the beam is obstructed. This is totally unacceptable for microwave systems. The Fresnel zone radius is given at each point by the following formula:

$$F_1 = 17.3 \sqrt{\frac{d_1 d_2}{fd}}$$

Where
F_1 = Fresnel zone radius in m
d_1 = distance of obstruction from one end in km
d_2 = distance of obstruction from other end in km
d = total path length ($d_1 + d_2$) in km
f = operating frequency in GHz.

By using $0.6F_1$ clearance on the microwave beam we are able to construct an effective system using the minimum height towers at each end. This is illustrated in Figure 3.20. The objective is to minimize the cost of both towers in order to minimize the total cost of system. However it may be preferable to use a taller tower at one site, say an existing TV mast, with a correspondingly smaller tower at the other site.

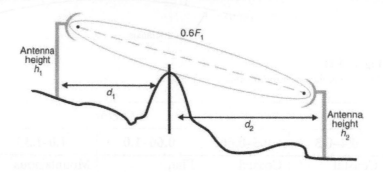

Figure 3.20
Fresnel zone clearance on microwave path

3.7.3   Radio link budget

Once the heights of the transmitting and receiving towers have been established the designer is in a position to select the appropriate antenna, or waveguide, transmitter power and receiver sensitivity to operate the proposed system. The next step is then to calculate the link budget to verify that the design will operate satisfactorily. To do this we need to check, on both of the link transmitting frequencies, that sufficient signal arrives at the chosen receiver. This is done by consulting the appropriate data sheets for the chosen items of equipment to determine their appropriate gains and losses. A schematic diagram of the overall system is illustrated in Figure 3.21.

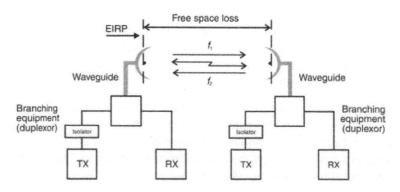

Figure 3.21
System schematic layout

The system link budget is then calculated using the following methodology:

- The free space loss along the radio path is calculated (L_{FS})
- The effective power produced by the transmitter is calculated (EIRP)
- The effective gain of the receiving antenna is obtained (G_{RX})
- The losses of all components in the receiving chain is calculated (L_{RX})

- The signal arriving at the receiver is the algebraic sum of all the above gains and losses and can be calculated as follows:

$$\text{Received signal} = \text{EIRP} - L_{\text{FS}} + G_{\text{RX}} - L_{\text{RX}} \text{ dBW}$$

- This can be compared with its receiver sensitivity limit (RSL).

Providing the received signal exceeds the RSL with sufficient fade margin then the system is deemed satisfactory.

Free-space loss

Free-space loss is a measure of how the transmitted energy expands out into space as it propagates between the transmitter and receiver. It is calculated from the following formula:

$$L = 92.4 + 20 \log f + 20 \log d$$

Where
  L = free-space loss in dB
  f = transmitted frequency in GHz
  d = path length in km.

Effective isotropically radiated power

Effective isotropically radiated power (EIRP) is a measure of the power effectively radiated from the transmitting antenna. It is calculated by taking the transmitter output power, subtracting from it the losses in the branching equipment and in the waveguide and then adding the effective gain of the transmitting antenna. This is illustrated in Figure 3.22.

$$\text{EIRP} = \text{TX power} - \text{TX losses} + \text{TX antenna gain}$$

Where
  EIRP is measured in dBW
  TX power is measured in dBW
  TX losses include all components in transmit chain in dB
  TX antenna gain is measured in dBi.

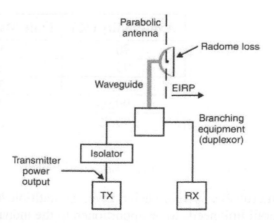

Figure 3.22
Components of EIRP

Microwave antenna

Microwave systems normally use a parabolic dish antenna, in which the small radiating element is placed at the focus of a paraboloid. This gives the antenna very high gain and a correspondingly narrow beamwidth. For example a 3.0 m diameter antenna may have a gain of 43 dBi at 6 GHz and a beamwidth of 0.5°. Such an antenna needs to be mounted on a very rigid structure to avoid the beam moving off the receiving antenna. Antenna gain and beamwidth figures can be obtained from manufacturer's data sheets. The following formula gives an approximation to the gain of a parabolic antenna:

$$G = 17.8 + 20 \log_{10} d + 20 \log_{10} f$$

Where

G is the gain in decibels relative to an isotropic antenna in dBi
d is the diameter of the antenna in m
f is the operating frequency in GHz.

Receiver sensitivity limit

The receiver sensitivity limit is the level of RF signal at the receiver input that will open the receiver, this is also referred as squelch. This is obtained from the receiver manufacturer's specification. A more suitable sensitivity level to use is the 12 dB SINAD sensitivity which is the RF input level required at the input to a receiver for a 1 kHz modulated wave (± 3 kHz deviation) that will give a 12 dB SNR at audio output level. This is also obtained from the receiver manufacturer's specification.

Fade margin

The fade margin is the amount of excess signal above the RSL. Because of the extreme variability of the radio path, it is possible for the signals level to drop by as much as 30 dB. It is therefore essential that the system has adequate fade margin to accommodate such variability. The amount of fade margin needed depends on the required availability of the link. When the bit error rate exceeds the system performance requirement the link is said to be unavailable. Availability percentages represent the percentage of the year at which the link is available. For example a 99.99% availability means that the system is unavailable for 0.01% of the year, or equivalently 547 s per year, or about 1.5 s per day (see Table 3.7).

Availability (%)	Fade Margin (dB)
90	10
99	20
99.9	30
99.99	40
99.999	50

Table 3.7
Indicates recommended fade margins for various availabilities

If the microwave system design involves multiple hops, and then the time availability of the overall link needs to be apportioned to the individual links. For example assume we require 99.99% availability for a system involving six links in series. The unavailability of the total system is then 1.0000–0.9999 or 0.0001 and this equates to 0.0000166 for each link. Therefore the availability per hop needs to be 1.0000–0.0000166 or 0.999983 that is 99.9983%.

3.7.4 Applications of microwave radio

Radio systems have a number of advantages plus a few disadvantages. The advantages of radio include:

- Systems only require a transmitting and a receiving site, with no requirement for a continuous right-of-way between them.
- Systems can be installed over inhospitable terrain, such as mountains or lakes, where laying of fiber-optic cables is impractical.
- Systems can be installed on mobile equipment.
- System security involves only the protection of the transmitting and receiving sites, without the need to protect cable along the total path length.
- System installation can be much quicker with only the two terminals to install.
- Systems are often installed as a backup to high capacity fiber-optic links, to ensure route diversity.

The disadvantages of using microwave radio systems are as follows:

- Radio systems have limited bandwidth, thereby restricting the achievable data rates.
- Radio signals are continuously being affected by changing atmospheric conditions, and multi-pathing effects.
- The radio medium is shared with other users and is susceptible to interference.
- The systems require licensing to control usage. Obtaining licences can incur administrative delays, and the licence conditions can restrict equipment operation.
- Only limited radio frequencies are available in certain areas.

3.8 Satellite

Satellite systems are another form of microwave radio transmission whereby the transmitter sends the signal up to the satellite, where it is amplified and then transmitted back to the earth on another frequency. This provides a point-to-multipoint transmission system where each terminal accesses its signal out of the common beam. The satellite is located in a geostationary earth orbit (GEO) at some 36 000 km above the earth's equator where it appears to be stationary relative to an observer on the earth. This is illustrated in Figure 3.23. This means that the satellite antennas at all of the terminals can have fixed alignment. In this section the principles of satellite operation and methods of access will be introduced and the design of very small aperture satellite (VSAT) systems will be discussed.

3.8.1 Satellite operating principles

Satellite electronics

The satellite comprises a number of transponders which receive the signal from the earth, amplify it, convert it to a lower frequency then amplify the signal for retransmission back to the earth. A block diagram showing these components is given in Figure 3.24. The low noise amplifier (LNA) is one of the most critical components in the satellite link since it is required to extract very low signals and the same time provide very large gain – up to

50 dB. The high power amplifier (HPA) in the satellite transponder may produce 10–50 W depending on the configuration. This power is limited by international agreement, to prevent interference with terrestrial microwave systems and such systems are said to be 'downlink limited'. By comparison, the equivalent HPA at the earth station may be running at power levels between 100 and 600 W.

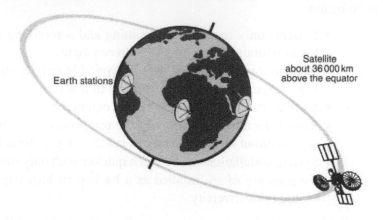

Figure 3.23
Geostationary earth orbit satellite communication

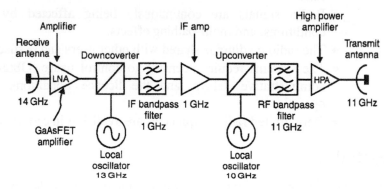

Figure 3.24
Block diagram of satellite transponder

Satellite frequency bands

The main frequency bands used for commercial satellite communication are listed in Table 3.8. It will be noted that the uplink frequencies, which are used from the earth to the satellite, are higher than the corresponding downlink frequencies. The reason for this is that the higher frequencies have a greater path loss and therefore require greater transmitter power. This is easily provided by the mains-powered earth stations, whereas the satellite transmitters are reliant on limited battery and solar energy resources, so use the lower frequencies.

Frequency Band Usage		Uplink Frequency (GHz)	Downlink Frequency (GHz)
L	Mobile	1.550–1.600	1.500–1.550
S	Military/government	5.925–6.055	2.535–2.655
C	Commercial	5.725–7.075	3.700–4.800

Frequency Band Usage		Uplink Frequency (GHz)	Downlink Frequency (GHz)
X	Military/government	7.900–8.400	7.250–7.750
Ku	Commercial	12.75–18.10	10.70–13.25
Ka	Military/government	27.00–31.00	18.30–22.2

Table 3.8
Satellite frequency bands

The reasons for using particular frequency bands are as follows:

- **C Band**

  - Most popular because least affected of all the bands by man-made and natural noise
  - Shared with terrestrial services, satellite transmit power is therefore limited to reduce interference.

- **Ku Band**

  - As C band is crowded Ku band is becoming most popular
  - Affected by rain attenuation and high natural noise levels
  - Greater data rates are possible with higher frequencies.

- **L Band**

  - Used for mobile low capacity terminals
  - Affected by man-made noise
  - Less pointing accuracy required
  - Lower power required.

- **S, X and Ka Bands**

  - Restricted to military, government and experimental users.

VSAT antenna diameter

The diameter of the antenna used at the earth station limits the amount of bandwidth that station is able to access. VSAT systems use antennae ranging from 5 to 11 m in diameter at the hub and at the remote terminal antenna diameter ranges from 0.5 (1.6 ft) to 2.5 m (8 ft). The outbound traffic from the hub may reach 2 Mbps, while the inbound traffic is typically 9600 bps.

Satellite access techniques

Satellite systems can be accessed in either the frequency domain (FDMA) or the time domain (TDMA). This access can also be assigned permanently (i.e. dedicated) for heavy traffic routes or dynamically as in demand-assigned multiple access (DAMA) systems. In DAMA systems the master station assigns traffic circuits to an earth station on demand. When the call terminates, the DAMA circuit is returned to the pool for reassignment to other users.

Frequency division multiple access

In frequency division multiple access (FDMA) systems each transponder, which typically has 36-MHz bandwidth, is assigned divided into various frequency segments. Each earth station is assigned one or more frequency segments. For example 14 earth stations could access one 36-MHz transponder with each having 24 analog voice channels. These allocations can be on a single channel per carrier (SCPC) basis whereby an individual user can transmit and receive on particular pre-allocated channels from his remote terminal. This is normally the simplest method of providing access to large numbers of small users.

The advantages in using FDMA are that no network timing is required and that the channel assignment is simple. The disadvantages are that uplink power levels need to be closely co-ordinated for efficient use of transmitter RF output power and intermodulation becomes a problem as the number of separate RF carriers increases.

Time division multiple access

Time division multiple access (TDMA) is used with digital transmission and shares the satellite transponder on a time basis. Individual time slots are assigned to the earth stations in sequence and each station has full and exclusive access to the whole of the transponder bandwidth during its assigned time slot. During its access period the earth station transmits a burst of digital information. All stations on the network need precise time synchronization. This is done using a synchronization burst at the start of each frame. A typical INTELSAT TDMA system has a frame period of 750 μs.

The advantages of TDMA are that only one earth station is providing input to the satellite at any one instant so intermodulation does not occur, the system allows greater flexibility with the different user's EIRP and data rates, and is able to cope quickly with the dynamically changing traffic patterns. The disadvantages of TDMA are that very accurate timing is required for all stations on the network, there is some loss of throughput due to the frame overhead and large buffer storage is needed to cope with long frame lengths.

3.8.2   Satellite system design concepts

Receiving system figure of merit (*G/T*)

The receiving system figure of merit (*G/T*) describes the capability of a satellite or earth station to receive a signal. This is defined as follows:

$$G/T = G - 10 \log T$$

Where
   G/T is the figure of Merit measured in dB per degree Kelvin (dB/K)
   G is the net antenna gain up to the reference plane, which is normally the input to LNA, in dB
   T is the effective noise temperature of the receiving system in degrees Kelvin.

Net antenna gain

The net antenna gain is the gross gain of the antenna less the losses of all of the components making up the downlink receive chain, namely the radome loss, feed loss, directional coupler loss, wave guide loss, the band pass filter loss and transition losses. These components are illustrated in Figure 3.25.

System noise temperature

The system noise temperature in degrees Kelvin is given by the following formula:

$$T_{sys} = T_{sky} + T_{feed} + T_{recv}$$

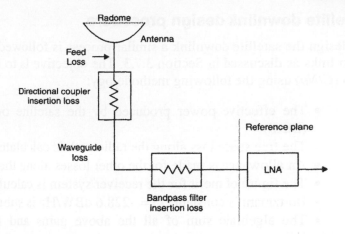

Figure 3.25
Components in receiver downlink chain

Where

T_{sky} is the sky noise
T_{feed} is the thermal noise generated by the receive chain components
T_{recv} is the thermal noise generated by the LNA.

Sky noise

Sky noise varies in proportion to frequency and inversely in proportion to the elevation angle. Some representative values of sky noise are given in Table 3.9. Earth stations generally have minimum elevation angles at which they will work satisfactorily. Below the minimum angles the radio beam is subject to much greater atmospheric refraction and there is greater noise pickup from terrestrial sources and the sky noise increases dramatically. The minimum elevation angles are 5° at 4 GHz and 10° at 12 GHz.

Frequency (GHz)	Elevation Angle (°)	Sky Noise (K)
4.0	5	28
4.0	10	16
7.5	5	33
7.5	10	18
11.7	10	23
11.7	15	18
20.0	10	118
20.0	15	100
20.0	20	80

Table 3.9
Some sky noise figures

Noise figure

The noise figure (*NF*) in dB is the thermal noise generated by the losses of all devices inserted in the receive chain and it can be related to the noise temperature by the following formula:

$$NF = 10 \log(1 + Te/290)$$

Where *Te* is the effective noise temperature measured in degrees Kelvin (K).

3.8.3   Satellite downlink design procedure

To design the satellite downlink a similar process is followed to the design of microwave radio links as discussed in Section 3.7.3. The objective is to calculate the carrier-to-noise ratio (*C/No*) using the following methodology:

- The effective power produced by the satellite or earth station is calculated (EIRP)
- The free space loss along the radio path is calculated (L_{FS})
- An allowance is made for the other losses along the radio path (L_{other})
- The figure of merit for the receiver system is calculated (*G/T*)
- Boltzmann's constant (k) = –228.6 dBW/Hz is subtracted
- The algebraic sum of all the above gains and losses can be calculated as follows:

$$C/No = \text{EIRP} - L_{FS} - L_{other} + G/T - k \text{ dB}$$

A fade margin of 4 dB is added to compensate for propagation anomalies and to allow for ageing in the components.

The other losses in the radio path can include:

- Polarization loss (0.5 dB).
- Pointing losses at both the earth station and the satellite (0.5 dB each). This is to compensate for any antenna misalignment.
- Off-contour loss to compensate for variation in received signal strength at the terminal location. This information is obtained from the satellite service provider in the form of an antenna footprint, which shows the downlink received signal strength (EIRP) for the specific spot or zone beam from the satellite. An example is shown for the Aussat 160E satellite in Figure 3.26.

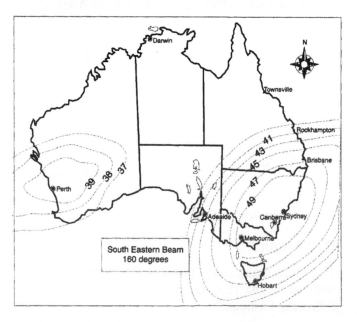

Figure 3.26
Example of satellite downlink footprint

- Gaseous absorption loss is caused by atmospheric absorption and increases with frequency, it is inversely proportional to the elevation angle, and the elevation of the earth station. At higher altitudes the air is less dense and therefore the path loss is reduced. The elevation angle is compensated for by allowing 0.5 dB for 4 GHz systems at less than 8° elevation and 1.0 dB for 7.25 GHz systems at less than 10° elevation.
- Rainfall attenuation is significant for systems operating above 10 GHz. The amount of excess attenuation is dependent on the rainfall intensity and the frequency.

3.8.4   Applications of satellite systems

Satellite systems have a number of advantages plus a few disadvantages. The advantages of satellite include:

- Systems only require a transmitting and a receiving site, and the distance between them on the earth is irrelevant.
- Systems can be installed in very remote locations where the provision of other services such as fiber-optic cables is impractical or prohibitively expensive.
- Systems can be installed on mobile equipment.
- System security involves only the protection of the transmitting and receiving sites, without the need to protect cable along the total path length.
- System installation can be much quicker with only the two terminals to install.
- VSAT systems are often installed to bypass the PSTN to create private data networks, ensuring data security. An example of a VSAT network topology is shown in Figure 3.27. These are often used for many outlying branches communicating back to a central site which could be head office or a warehousing center, etc.

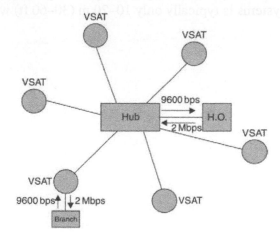

Figure 3.27
Typical VSAT network configuration

The disadvantages of using satellite systems are as follows:

- Satellite systems have limited bandwidth, thereby restricting the achievable data rates.
- The radio path is very long, and receive power is very low making systems susceptible to interference.

- The length of the radio path, some 36 000 km each way, causes a propogation delay of about 270 ms.
- Only limited radio frequencies are available in certain areas.
- GEO satellite orbital positions are limited and radio frequency interference is possible if an earth station antenna is incorrectly aligned or has too wide a beamwidth.

3.9   Infra-red

Infra-red (IR) systems are frequently used for local area networks using radiated light of about 800–900 nm wavelength. These systems do not require licensing and are immune to radio frequency (RF) interference. However they are susceptible to interference from the Sun or incandescent or fluorescent light fittings. IR receiver design is relatively simple as such systems normally modulate the signal amplitude rather than its frequency or phase to convey information. This makes the equipment cheaper than FM units.

Two approaches are used with infra-red local area networks. The signals can either be focused and aimed or used for omnidirectional transmission. Focused infra-red systems are only suitable for fixed terminals and can achieve much higher signal-to-noise ratios and have fewer problems with multipath interference. Such systems can be used outdoors with ranges up to several kilometers. An example of this technology is a fully compliant IEEE 802.5 token ring system operating at either 4 or 16 Mbps.

The alternative approach is to use omnidirectional transmission. Here the transmitters are either mounted on the ceiling or the signals bounced off the ceiling to provide coverage within a room. Many building materials are opaque to IR radiation so the line of sight paths are generally required. White plasterboard reflects IR signals so coverage to the less accessible parts of the room can be achieved by reflecting the signals off the walls or ceiling. However this can also cause multipath interference. However this can often be even rejected by suitably aiming the receiver. The coverage of omnidirectional IR systems is typically only 10–20 m (30–60 ft) within one room.

4

The public switched telephone network (PSTN)

Objectives

When you have completed study of this chapter you should be able to:

- Describe the structure of the PSTN
- Discuss the factors limiting the length of subscriber loop plant
- Discuss the role of concentration and expansion in a local switch
- List three basic functions of a local switch
- Explain the basic principles of time division switching
- Discuss the advantages of common control switches
- List the functions of a subscriber's line circuit
- Describe the function of a hybrid transformer
- Illustrate the use of compelled signaling sequences
- Discuss the advantages of common channel signaling
- List the layers used with SS No. 7.

4.1   Introduction

In this chapter the structure of the public switched telephone network (PSTN) will be investigated, together with the important infrastructure of the CCITT Signaling System No. 7. The PSTN provides the communications infrastructure to carry voice and data locally, nationally or internationally throughout the world. It has evolved from being an analog, voice-only, communications network to a sophisticated high-speed digital network capable of carrying any form of data communications, including digitized speech. The PSTN is made up of local networks connected by a long-distance network as shown in Figure 4.1. The PSTN is used to provide connections between end-users. These are defined by the IEEE as 'an association of channels, switching systems, and other functional units set up to provide means for a transfer of information between one or more points in a telecommunications network'. The local network comprises the end-users and circuits connecting them to local switching points called nodes. The nodes switch the communications traffic from the originating local network to the destination local network by means of trunks.

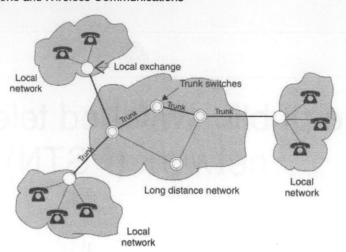

Figure 4.1
PSTN overview

Because open access to the PSTN is vital for commerce it is normally regulated by a government authority. In many countries the PSTN is operated as a government monopoly but there is an increasing trend towards privatization and increased competition, particularly in the area of long-distance service provision. PSTN circuits are increasingly used to carry data, particularly for Internet access. Some technologies enable all or part of the PSTN to be bypassed. These include private networks operated by private service providers or utilities, such as railway or power companies and satellite systems enabling direct customer-to-customer communication.

4.2   Local network

The local network comprises the subscriber plant plus the switching infrastructure at the node. The subscriber plant consists of the appropriate terminal equipment to interface to the distribution network connecting to the local switch. The network interfacing equipment is normally called the customer premises equipment (CPE), while the distribution network normally comprises a subscriber loop of twisted-pair copper. Most of the traditional local networks preferably provide, at least two dedicated, individual copper pairs for each-end user. More than 65% of the investment in a PSTN is in the local loop, so more efficient methods are being applied to provide the required bandwidth for the customers. With the advent of modern electronic devices it is becoming cost-effective to utilize some form of concentrator to consolidate multiple customers onto one circuit. Such concentrators may also incorporate local switching so that the two neighbors do not need to utilize equipment at the central switch for their 'chat over the back fence'. By making the conventional twisted-pair copper subscribers loops shorter, they are able to support higher bit rates using some of the digital subscriber line techniques, such as ISDN described in Chapter 6 or alternatively ADSL or EtherLoop as detailed in Chapter 7.

4.3   Subscriber plant

4.3.1   Subscriber loop

The subscriber plant has conventionally used twisted-pair copper. This provides the DC power to operate the basic telephone instrument from the Central Office, conveys the AC ringing voltage used to operate the telephone bell to indicate incoming calls and conveys loop signaling to the Central Office so as to indicate when the telephone is on- or off-hook. This is used to indicate when the customer wishes to initiate a call, by going off-hook or to terminate a call by replacing the telephone handset; that is going on-hook. The metallic circuit can also be used to convey loop pulsing from the old style rotary telephone dial. As the telephone dial rotated it caused the DC current in the subscribers loop to be made and broken at about 10 pulses per second to indicate the number of the required called party. This is called loop decadic signaling, but has largely been replaced by the use of touch-tone signaling in which two-tones are simultaneously transmitted for each key pressed on the subscriber's key pad. Touch-tone signaling has the advantage of not requiring a metallic circuit to operate.

The factors which limit the subscriber loop length are attenuation and resistance. The attenuation in the subscriber plant needs to be limited to ensure that adequate signal levels are maintained at the telephone instrument. The maximum loss limit is measured at a reference frequency. In North America this is 1000 Hz whereas 800 Hz is generally used elsewhere in the world. The maximum loss objective for a subscriber loop is 8 dB in North America and may be as low as 7 dB in another countries.

The loop resistance needs to be limited to ensure sufficient current flows in the subscriber's loop to operate both the telephone instrument and the loop detector at the local switch. The minimum loop current is generally 20 mA which corresponds to a loop resistance of 1700 Ω. This was discussed in Section 3.3.4.

4.3.2   Copper subscriber loop plant

The conventional subscriber distribution plant is made up of twisted-pair copper. The diameter of the copper wire determines its resistance and distribution networks commonly comprise either 0.5 mm (24 AWG) or 0.63 mm (22 AWG) conductors. Access points are designed into the network in the form of cross connection cabinets to give more flexibility in the network and provide cost-effective circuit provision. The dedicated distribution cable to the customer's premises is generously provisioned to allow for future growth and maintenance spares. This distribution cable terminates at the cross connection cabinet where it is connected by means of a jumper wire to a feeder cable back to the Central Office. This is illustrated in Figure 4.2.

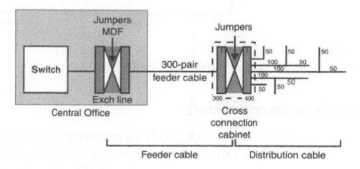

Figure 4.2
Conventional subscriber loop plant configuration

The distribution cable can be dedicated for individual subscribers, and is frequently designed on the basis of two cable pairs per residence. The distribution cable is often installed in 50-pair increments. If a dedicated cabling system is used, the remainder of each pair in the 50-pair cable is cut off beyond the dedicated residence to prevent its capacitance shunting the working circuit. If this is not done a bridged tap is created as shown in Figure 4.3.

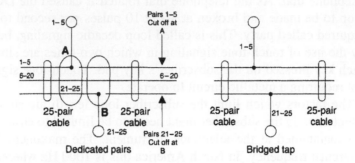

Figure 4.3
Dedicated pairs and bridged taps

The effect of such taps is to absorb certain frequencies, thereby limiting the ability of such pairs to carry high frequency transmission systems such as ADSL. A typical frequency response curve for a cable with a bridged tap is shown in Figure 4.4.

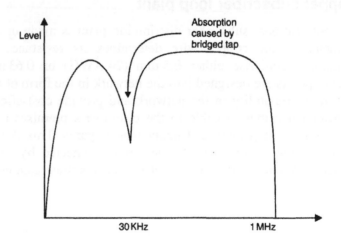

Figure 4.4
Frequency response curve for cable with a bridged tap

4.3.3   Remote subscriber unit

The remote subscriber unit (RSU) is increasingly being used to connect the distribution cable to the Central Office. It provides the subscriber line interface card (SLIC) for conventional plain old telephone service (POTS). This is discussed in Section 4.5. The RSU can also carry out a concentrator function with an optional local switching capacity. The RSU may also provide independent interfaces for high-speed data circuits. The RSU is particularly useful for remote customers who would require the use of derived feeder circuits, such as radio, a fiber-optic cable or a digital subscriber line, in order to meet the objective attenuation limits as set out in the national transmission plan. Some alternative RSU configurations are shown in Figure 4.5.

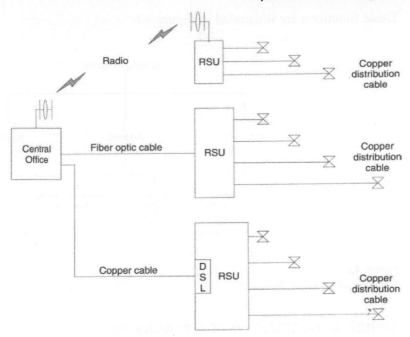

Figure 4.5
Alternative RSU configurations

4.4 Telephone switching infrastructure

4.4.1 Switching overview

Telephone exchange switching enables customers on any local exchange to be connected to any other subscriber in the world simply by dialing the appropriate digits on their telephone instrument. The switching network analyzes the digits dialled and sets up an appropriate connection to the called number which could be in the next street or on the other side of the world. To do this the switch requires some form of 'intelligence'.

The basic functions of the switch are as follows:

- The switch must be able to connect any incoming call to any one of its subscribers or outgoing circuits to other switches.
- The switch must establish and hold the physical connection across the network between the caller and the called subscriber for the duration of the call.
- The switch must monitor the status of the lines of both parties to the call to determine when either party has ended the conversation by replacing their handset. When this condition is detected the switch must clear the physical network connection to allow the equipment to be re-used for another call.
- The switch must prevent any new calls intruding into the connections that are already in use.
- When the equipment in the switch is busy or unavailable, the caller must receive appropriate supervisory tones such as 'busy tone' or 'overflow busy tone' or suitable voice announcements.

The signaling system has three main functions:

1. To tell the various system components what to do next
2. To inform the subscribers of the progress of the call
3. To initiate and terminate the billing process.

These functions are illustrated in Figure 4.6.

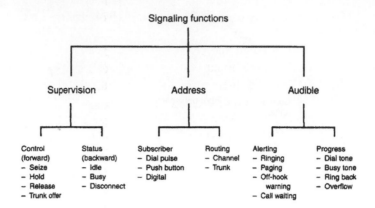

Figure 4.6
Summary of signaling functions

4.4.2   Switch concentration and expansion

The local switches utilize concentration whereby all of the individual subscribers are able to access a limited number of simultaneous switching paths to other subscribers or trunks to other exchanges, and expansion enabling a call on one of those limited paths or trunks to be connected to any one of the called subscribers. The concentration ratio is the same as the expansion ratio and may be in the order of 10:1 for residential subscribers or 3:1 for business subscribers. Switch design is based on the appropriate grade of service for that kind of customer and may assume, for example, that during the exchange busy hour on average only 10% of the customers will require to make calls simultaneously. If the number of customers wanting to make calls exceeds the number of switch paths or trunks available at that time, then the extra customers are temporarily denied service and receive a tone or voice-announcement indicating the congestion condition. This results in a more cost-efficient exchange design than providing each subscriber with their own switch path. In the worst case scenario, with half the customers talking to the other half, the number of connections is only 50% of the number of customers. Figure 4.7 indicates the concentration and expansion stages of a local switch.

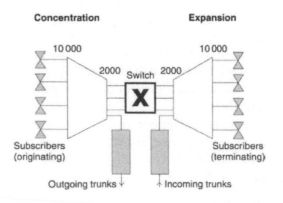

Figure 4.7
Concentration and expansion stages of a local switch

4.4.3   Space division switching

Space division switching involves the physical connection of one path to another. This kind of switching was used in all the earliest exchanges, and is still an essential part of modern digital exchanges.

In a manual exchange the operator plugged a cord into the calling subscriber's line appearance on her switchboard to answer the call and ascertain the number of the called subscriber. She would then plug the associated cord into the called subscriber's line appearance to complete the call switching.

The automatic step-by-step (SXS) or Strowger switch was based on electromechanical switch banks in which the switch would step vertically, one level in response to each pulse in the digits dialed. This used progressive control where the digits dialed by the subscriber's telephone operate each switch in turn. In the delay period between digits, the next switch in the chosen path would be prepared to receive the next dialed digit. In this way a physical path was progressively extended from the calling subscriber to the called number.

A later version of space division switching is used in crossbar switches. Crossbar switches are set up as a matrix in which the simultaneous operation of a horizontal relay and a vertical relay operate a pair of contacts to complete the speech path. These contacts then latch to maintain the speech path for the duration of the call, that is until the 'on-hook' condition is detected.

The modern digital exchanges use space division switching to connect different digital data streams in a spatial domain. These normally utilize logic gates as the switching elements, arranged in a matrix so that a horizontal input stream can exit on any one of the vertical output streams. This is illustrated in Figure 4.8, which shows a matrix having M horizontal inputs and N vertical outputs. This is called an $M \times N$ array and if $M = N$ the switch is non-blocking, since every input can be connected to an outlet, while if $M > N$ the switch has a concentrator action and conversely if $M < N$ the switch has an expansion role.

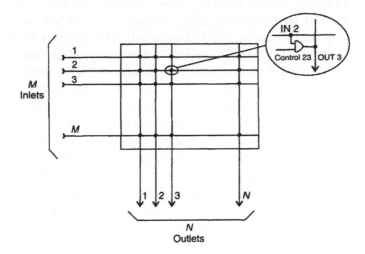

Figure 4.8
$M \times N$ *array of logic gates*

4.4.4   Time division switching

Time division switching is accomplished by interchanging time slots between digital data streams. You will recall that in digital systems each speech channel is sampled at

8000 times per second with each 8-bit sample placed in a time slot. Time switching involves storage of the data from the various incoming time slots then placing that data in a different sequence in the outgoing time slots. This is shown diagrammatically in Figure 4.9.

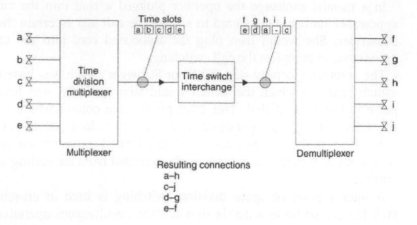

Figure 4.9
Time slot interchange

In this simple example the time division multiplexer on the left samples each of the customers 'a' to 'e' in turn producing the sequence of five time slots 'abcde'. This enters the time switch where the time slot samples are rearranged to emerge as the sequence 'eda-c'. These time slots correspond to customers 'fghij' when demultiplexed on the right of the drawing. As a consequence the time slot data from customer 'a' is sent to customer 'h'. Likewise connections are made between c–j, d–g and e–f. Note that the data from customer 'b' was not sent out of the time switch, and the time slot for customer '*I*' in this case was empty.

In practical digital switches large numbers of time slots are used within each time switch, and two separate paths through the switches are simultaneously used for the outgoing and incoming parts of each call. The switches are usually designed in multiples of 120 channels which fits well with five groups of 24 channels from the North American (T1) transmission systems and four groups of the 30 channel European (E1) transmission systems. Modules of 120 telephone channels would probably require 128 time slots and if these are switched at a bit rate of 64 kbps then the switch highway speed corresponds to a bit rate of 8.192 Mbps. To increase the number of time slots most manufacturers utilize parallel switching within their time switches, operating the high-speed highways in an 8-bit parallel format. In this way a highway providing 528 time slots and 480 telephone channels can be operated at an effective bit rate of 4.224 Mbps.

Practical digital switches are built up of combinations of time-division (T) and space-division (S) switches. This allows the time slots from one highway to be interchanged with slots on a different highway thereby greatly expanding the size of the switch. A switch comprising TSST switching stages is illustrated in Figure 4.10.

The functions of the different stages in this example TSST switch are as follows:

- *T1 (time switch 1)*:   Shifts data from one time slot to another on the same highway entering the space switch.
- *S1 (space switch 1)*:   Shifts time slots from one of the incoming highways to one of the junctor highways to space switch 2, without changing the time slot positions.

- *S2 (space switch 2)*: Shifts time slots from one of the junctor highways to one of the highways to time switch 2, without changing the time slot positions.
- *T2 (time switch 2)*: Shifts data from one time slot to another on the same highway leaving the space switch.

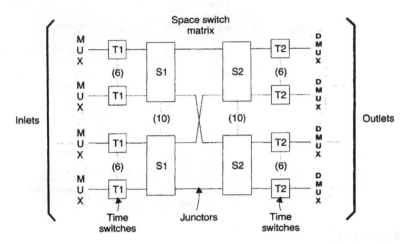

Figure 4.10
Example of TSST switch

4.4.5 Call processing control

There are two fundamental types of switching control methods used in telephone exchange systems – direct progressive control and common control.

Direct progressive control systems such as step-by-step exchanges are controlled directly by the telephone dial. As each digit is dialed the switch, called a group selector, is stepped upward in accordance with each pulse to select the appropriate digit level. During the interdigit pause the group selector then hunts for an outlet (1 out of 20) to a free selector to process the next digit. When the penultimate digit has been dialled the final selector steps up to the appropriate level and awaits the receipt of the final digit. As this is received it steps on each impulse around to the appropriate outlet corresponding to the equipment address of the called subscriber. The resultant post dialing delay is minimal since connection to the called subscriber takes place immediately after the digit is dialed. However the routing in such exchanges is fixed by the numbering scheme and is inflexible.

All modern exchanges such as crossbar and stored program control (SPC) exchanges use common control systems which are indirectly controlled. In the crossbar systems the control is undertaken by a number of specialized processors called markers whereas in a stored program control systems all processing is undertaken by a centralized computer system.

In the crossbar exchange the calling subscriber is connected to a free register by the marker. Having quickly made this connection the marker is free to undertake other tasks. The register receives and stores the dialed digits from the subscriber and when the dialing is completed, the marker returns and connects the subscriber to the appropriate outgoing circuit or called subscriber. This it can be done by using any appropriate path through the switching system, regardless of the actual digits dialed. A translator is used to determine the appropriate route to the destination independent of the numbering scheme. The numbers of the various items of common control equipment, such as markers and registers, required to process the calls depends on the number of subscribers and the traffic they generate. A simplified common control crossbar system is illustrated in Figure 4.11.

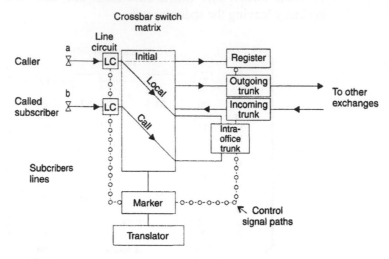

Figure 4.11
Common control crossbar system

Stored program control (SPC) systems operate in a similar manner to the crossbar system described above except that the logical functions of the register, marker and translator are all contained within the central processor and controlled by its software.

4.4.6   Call supervision

The call supervision process is undertaken continuously by the line circuit monitoring the state of the subscriber's loop. Since every working subscriber's line is connected to its own line circuit, these need to be checked continuously to ascertain whenever the subscriber has just gone 'off-hook' and wishes to initiate a call. Similarly when the subscriber has just replaced his handset, that is gone 'on-hook', the call requires to be terminated. This process is detailed in Section 4.6.5. The other supervision activity involves the tripping of ringing when the called subscriber picks up the handset to answer the telephone. This 'off-hook' condition signals to the exchange that the ringing voltage should now be turned off. This process is also carried out by the line circuit, as detailed in Section 4.6.4.

4.5   Local switches

4.5.1   Local switch overview

A local exchange or Central Office switch has inlets to serve incoming calls and outlets to serve outgoing calls. A call from the calling subscriber enters the switch via an inlet and is then connected to the calling subscriber on the outlet side of the switch. The switching function can be space-switched or time-switched or use combinations of the two. The switching function is initiated by the calling subscriber using line signaling conditions of either 'off-hook' or 'on-hook' together with dialing information to indicate the number of the called subscriber. The switch provides line supervision, processes the dialing information to determine the correct route for the call, interconnects the appropriate inlet and outlet and communicates with the next switch along the required route. The components of a local switch are illustrated in Figure 4.12.

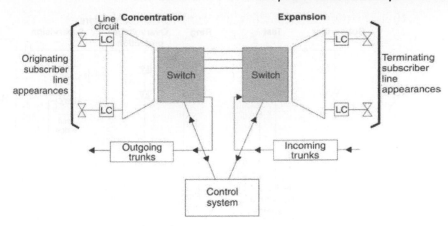

Figure 4.12
Central Office switch components

4.6 Subscriber line circuit

4.6.1 Line circuit functions

Each subscriber is connected to an individual line circuit in the Central Office. Depending on the type of switching system used there may be different types of line circuits used to handle different types of subscribers line: individual, 2-party, multi-party, PBX, DDI and coin box, etc. The line circuit in a modern electronic exchange carries out the following functions described by the mnemonic BORSHT:

B: Battery supply to the subscribers line
O: Over-voltage protection
R: Ringing current supply to the telephone
S: Supervision of the line
H: Hybrid for 2-wire to 4-wire conversion
T: Test access to the 2-wire line from centralized test equipment.

The line circuit also has a function to identify the calling subscriber for centralized billing purposes.

These functions are detailed in the following sections.

An outline of a subscriber's line circuit is shown in Figure 4.13.

Battery supply for subscriber's line

The line circuit uses a feed-bridge to supply the exchange battery voltage to the subscriber's line. This battery voltage is used to operate the conventional telephone instruments, and is used for the subscriber line signaling to identify whether the telephone instrument is 'on-hook' or 'off-hook'. The battery supply needs to be coupled through a high-impedance to the line, to prevent the low impedance of the battery shunting the speech. This task is done by means of a feed-bridge and usually involves two high impedance relay coils, providing supervision of the subscriber's loop. The relay will be operated by a low line impedance and released by a high impedance from the subscriber. This is illustrated in Figure 4.14.

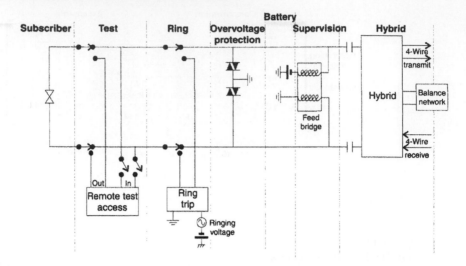

Figure 4.13
Outline of subscriber's line circuit

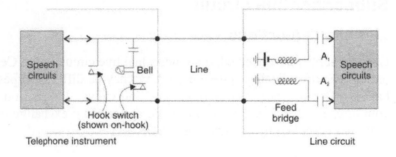

Figure 4.14
Subscriber's line feed-bridge

Overvoltage protection

The subscriber's line circuit in modern telecommunications equipment uses proprietary integrated circuits and as such is very susceptible to excessive voltages from the subscriber's line. Suitable over voltage protection is therefore incorporated in the subscriber line circuits. This would often take a form of zener diode clamping along with a MOV over voltage protection. It should be noted that in areas of high lightning activity for example, the primary over voltage protection would be in the form of gas discharge tubes located on the main distribution frame (MDF). The location of the over voltage protection is clearly marked in Figure 4.13.

Ringing current supply

The current required to operate the subscriber's telephone alerting device to indicate incoming calls is supplied through the subscriber's line circuit. This is alternating current with the appropriate National frequency and ringing cadence(s) and can be as high as 140 V depending on the line length. This is AC coupled to the ringing detector in the telephone, so it does not affect the loop supervision. The ringing current is connected to the line through a ring trip circuit as shown in Figure 4.13. This circuit is designed to remove the ringing voltage from the line as soon as the subscriber answers the telephone by taking the receiver off-hook.

Line supervision

Supervision of the subscriber's loop needs to be maintained continuously to determine whether the subscriber's line is in the 'on-hook' or 'off-hook' condition. The subscriber's loop transition indicates whether the subscriber wishes to originate or terminate the call. As discussed in Section 4.6.2, this supervision is typically maintained from the relay in the feed-bridge.

Hybrid transformer

In digital exchanges a hybrid transformer is used to convert the 2-wire subscriber's loop into a 4-wire circuit, in which the speech is separated into transmit and receive paths. The hybrid is basically a 4-port, 4-winding transformer which splits the power between four sets of 2-wire circuits as shown in Figure 4.15. These ports are respectively the 2-wire line, the 4-wire transmit path, the 2-wire hybrid balance network and the 4-wire receive path. The hybrid balance network is designed to match the impedance of the 2-wire line over the frequency range of the telephone channel. This is typically either a resistance-capacitance network made up of a 900 Ω resistance and a 2.16 μF capacitor in series to provide a 600 Ω impedance at 1000 Hz, or simply a 600 Ω resistance, but may need to be adjusted for very short lines, or non-standard cable impedance.

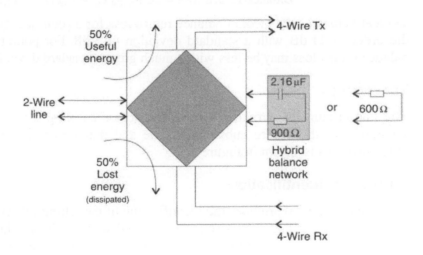

Figure 4.15
Basic hybrid transformer operation

The signal from the 2-wire line is divided equally between the 4-wire transmit and the 4-wire receive impedances, with the energy in the 4-wire receive path being dissipated as heat. Similarly the signal arriving from the 4-wire receive path is divided equally between the 2-wire line and the balance network impedances. In an ideal hybrid half of the signal energy goes to its intended destination while the remainder is lost. The hybrid would therefore have a dissipation loss of 3.0 dB, to which the typical insertion loss of the hybrid of about 0.5 dB needs to be added.

If the balance network does not correctly match the impedance of the receiving subscriber's line to which the talker was connected then some of the energy from the 4-wire receive path will leak across the hybrid into the 4-wire transmit path and be returned to the talker as echo. This is illustrated for a long-distance connection in Figure 4.16.

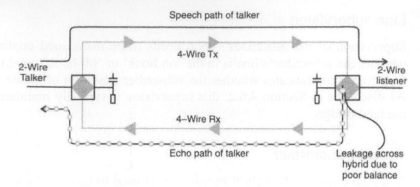

Figure 4.16
Echo in a long-distance connection

The extent to which this impairment disrupts the circuit depends on the relative magnitude of the echo signal, which is a function of the degree of impedance mismatch, and the amount of time the signal is delayed. The effect is subjectively worse on long-distance calls. Echo can be extremely disruptive to data circuits. The extent to which the impedance matching is achieved between a line of impedance Z_L and a balance network of impedance Z_N is defined as the balance return loss.

$$\text{Balance return loss} = 20 \log_{10}(Z_L + Z_N)/(Z_L - Z_N) \text{ dB}$$

Typical values of the median balance return loss for a good quality network would be of the order of 11 dB with a standard deviation of 3 dB. For poorer networks the median balance return loss may be less with a much greater standard deviation.

Test access

The line circuit also provides remote test access enabling service personnel to connect separately to the 2-wire subscriber's line and the subscriber's line circuit for testing purposes. This is shown in Figure 4.13.

Subscriber identification

The line circuit also provides the identification of the calling subscriber when required for any centralized billing purposes, such as long-distance calls or value added services, such as −900 calls. This information is also required where calling number identification is operating.

4.7   Trunk switching

4.7.1   Overview

The PSTN uses trunk switching at tandem or toll exchanges to interconnect the various local exchanges in the network. Trunk switching uses 4-wire connections between the local exchanges via any other tandem switching points between them.

International exchanges are also used to connect to other countries. To be able to conduct a satisfactory telephone conversation to another country it is important that both of the telephone networks comply with the appropriate technical recommendations of the ITU. Since half the telephone call is handled in the originating country and the other half in the destination country, satisfactory performance depends on both parties complying with the relevant recommendations. The most important of these cover the transmission performance and the signaling conventions.

4.7.2 Transmission quality of service

ITU-T recommendation E.171 specifies that an international connection shall have no more than 12 links in tandem. These links are apportioned so that four links are in the calling party's country, four links are used in the international connection and four links are in the called party's country. This then requires the design of national networks having no more than four links in tandem, to ensure the networks meet the transmission quality of service objectives. This is shown diagrammatically in Figure 4.17. The transmission quality deteriorates when more links are added in tandem. Post dialing delay increases when more tandem switches are added, the transmission impairments of bit error rates, pulse jitter and wander increase as more digital systems are connected serially in the transmission path.

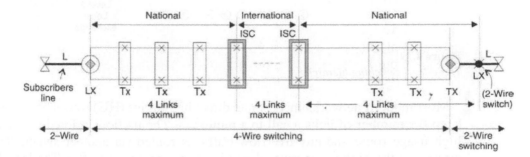

Figure 4.17
Link allocation for an international connection

4.7.3 Tandem switch

Long-distance networks use tandem or transit switches to connect between the local exchanges on the PSTN. These differ from local switches in two main respects; all trunk switching is 4-wire and trunk switches are non-blocking and neither compress nor expand the traffic since they have the same number of trunk inlets as outlets. The basic configuration of a tandem switch is shown in Figure 4.18.

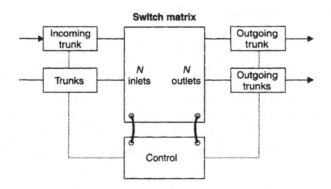

Figure 4.18
Basic configuration of a tandem switch

4.7.4 Exchange hierarchy

An important national network design decision is to determine the number of hierarchical levels in a national switching network. The top level of the exchange hierarchy is the international switching center (ISC) and more than one of these may be necessary.

The next level provides the long-distance switching using transit or toll exchanges while the bottom level would consist of the local exchanges and tandem exchanges to interconnect them. The objective of this hierarchy is to ensure that there are no more than four links in tandem on any connection between the local exchange and the ISC to ensure compliance with ITU-T recommendation E. 171 as discussed in Section 4.6.2. This concept is illustrated in Figure 4.19.

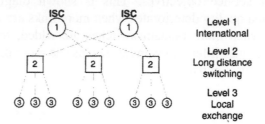

Figure 4.19
Three-level switching hierarchy

Most modern networks incorporate direct high usage (HU) routes between exchanges to keep the number of links a call to a minimum. The traffic is offered initially to the direct high usage route and any overflow traffic is routed via another switch. The number of circuits in the high usage route depends on the amount of traffic. The last choice route is called the final route and traffic from this route is not allowed to overflow into any other routes. An example is given in Figure 4.20, where traffic from A to B is offered initially to the direct high usage (HU) route. Any traffic overflowing from the HU route is offered to the final route via trunk exchange C.

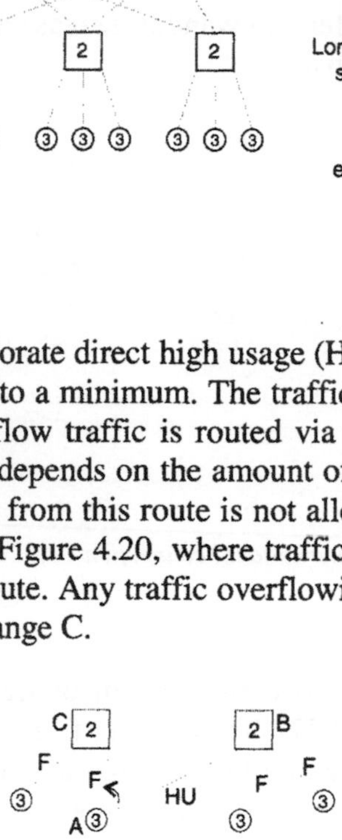

Figure 4.20
High usage and final routes

Dynamic routing is normally utilized in modern switches where the chosen route for a particular type of call attempt can be automatically varied to reflect the state of the network. These changes may be time-dependent, with the changes made at fixed pre-arranged times to accommodate changing traffic patterns. Adaptive routing schemes are state-dependent and change the routing patterns in accordance with changes in the state of the network. This is often done by a central routing processor which collects data on a regular basis from all the individual exchanges in the network. If the analysis of this data indicates the necessity for changing the routing patterns, then the required routing directives are sent to the appropriate exchanges.

4.7.5   Channel associated information signaling

Inter-exchange information signals transfer the address or telephone number of the called subscriber plus other call progress and control signals around the network. These can include, amongst others, the type of subscriber, the type of call, automatic number identification requests, clear forward and clear back signals, re-seize and re-answer signals in response to a hook-switch flash, and trunk offer signals whereby an assistance operator can gain access to busy lines. The early exchanges used channel associated signaling such as decadic signaling, and compelled multi-frequency code (MFC) signaling in accordance with CCITT recommendation R2.

Using R2 signaling, the signals are sent in the speech band using a combination of 2 out of 5 (or 2 out of 6) frequencies using tones in the range of 1380–1980 Hz for the forward direction and 540–1140 Hz in the backward direction. The use of two simultaneous tones minimizes the risk of false signaling. An example of MFC end-to-end signaling in the trunk network is shown in Figure 4.21. Here control of the call is passed to the toll exchange and two end-to-end signaling paths are set up; first between the originating exchange and the toll exchange and then between the toll exchange and the terminating exchange. As the call is extended from one tandem exchange to the next, the speech path is immediately switched and that incoming register released. In this way the originating register progressively passes subsequent information on to the incoming register in the next exchange, until the terminating exchange is reached.

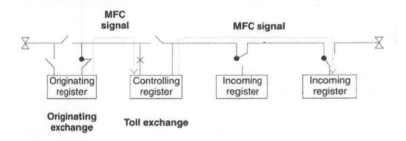

Figure 4.21
MFC end-to-end signaling

This signaling process is compelled and operates as follows:

- On seizure of the outgoing circuit the outgoing register starts sending the first forward signal.
- When the incoming register recognizes this signal it starts sending a backward signal as an acknowledgment.
- When the outgoing register recognizes the acknowledgment it stops sending the forward signal.
- When the incoming register recognizes that the forward signal has stopped it stops sending its backward signal.
- When the outgoing register recognizes that the backward signal has ceased it will then commence sending the next forward signal.

This process is illustrated in Figure 4.22. This method of signaling achieves high reliability but is comparatively slow.

4.7.6 Common channel signaling

With modern stored program controlled exchanges it is inefficient for the digital processor to handle channel associated signaling. A much more efficient method is to transfer all the signaling information between the processors over a high-speed data link. This is called common channel signaling (CCS). This has many advantages including having a signaling path which is entirely separate from the switched speech paths, very fast signaling is possible because of the wide bandwidth (64 kbps) of the signaling channel, and it enables provision of value-added services such as calling number identification, –800 and –900 services, etc. The principle of common channel signaling is illustrated in Figure 4.23.

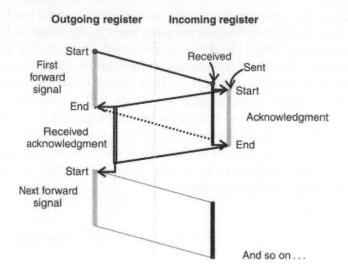

Figure 4.22
Compelled signaling sequence

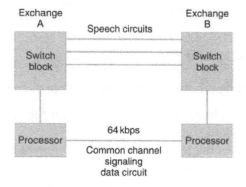

Figure 4.23
Common channel signaling

4.8   CCITT Signaling System No. 7

4.8.1   Overview

CCITT Signaling System No. 7 (SS No. 7) is the most commonly used signaling system for all digital networks using 64 kbps channels. It is simply a digital data network dedicated entirely to inter-switch signaling between stored program control (SPC) exchanges. The overall objective of SS No. 7 is to provide an internationally standardized general purpose common channel signaling system to meet the requirements of information transfer for inter-processor transactions with the digital communications networks for call control, remote control, network database access and management, and maintenance signaling. It provides a reliable means of information transfer in the correct sequence without loss or duplication. The provision of ISDN services (Chapter 6) is made possible by the existence of the SS No. 7 data communications system.

4.8.2 Signaling system structure

The signaling system utilizes the bottom four of the seven layers of the OSI model in order to minimize the post-dialing delay. ITU-T Rec.Q.709 specifies that the post-dialing delay shall be no more than 2.24 s for 95% of calls. To achieve this imposes a limit on the number of signal transfer points (STP) through which the signaling message has to travel. Network signaling plans are required to ensure compliance with the performance objectives as set out in ITU-T Rec. Q.706 specifying parameters for message delay, signaling traffic load and error rates. The worldwide signaling network has two functionally independent levels: international and national.

The basic structure of SS No. 7 is shown in Figure 4.24.

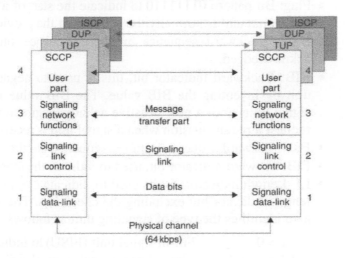

Figure 4.24
General structure of SS No. 7 layers

The message transfer part (MTP) embraces the bottom three layers of OSI model and four user part layers sit on top of it; the signal connection control part (SCCP) then the specific user parts for the relevant services: telephone (TUP), data (DUP) and ISDN (ISUP). An outline of the functions of each of these layers will be given in the following sections.

Signaling data-link layer (layer 1)

The signaling data link function defines the physical, electrical and a functional characteristics of the data link and the method of accessing it. This carries the data bits and in a digital network normally comprises a 64 kbps digital path, although the minimum bit rate for telephone call control is 4.8 kbps. This is a bidirectional transmission path using two data channels operating together in opposite directions at the same data rate, dedicated exclusively for the signaling information. Time slot 16 is normally used for signaling on a 2.048 Mbps (E1) digital link. This signaling data link may be accessed via a switched link allowing automatic re-configuration of the signaling data links.

Signaling link layer (layer 2)

The signaling link functions correspond to the data link layer of the OSI model. These define the method of transferring signaling messages over one signaling data link. They define the frame format of the signaling units, synchronization, flow control, error detection and correction procedures plus link failure detection and recovery procedures.

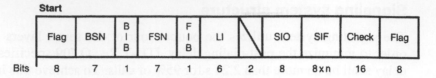

Figure 4.25
Message signal unit (MSU) format

The basic format of a message signal unit (MSU) is shown in Figure 4.25. The function of each of the fields in this message is as follows:

- Flag: Bit pattern 01111110 to indicate the start of a signal unit. The start flag of one signal unit is normally the end flag of the previous signal unit.
- BSN: Backward sequence number, sequence number of signal unit being acknowledged.
- BIB: Backward indicator bit, this is used to negatively acknowledge a signal unit by inventing the BIB value. The BIB value is maintained in subsequent signal units until a new negative acknowledgment is to be sent. This is used to request a retransmission when a signal unit is received out of sequence.
- FSN: Forward sequence number, sequence number of signal unit being carried.
- FIB: Forward indicator bit, used to indicate the start of a retransmission cycle.
- LI: Length indicator, 6-bits used to indicate the number of octets following the length indicator but excluding the check bits. The value of the length indicator also identifies the type of signaling unit as follows:

  LI = 0   Fill-in signal unit (FISU) to indicate link is in idle state
  LI = 1 or 2   Link status signal unit (LSSU) – provides link status information
  LI > 2   Message signal unit (MSU).

- SIO: Service information octet, divided into two 4-bit fields, the service indicator and the sub-service field. The service indicator is used to define the type of service or message, for example 0011 indicates the SCCP, 0100 the telephone user part, 0101 the ISDN user part and 0110 the data user part. The sub-service field is used for national/international message discrimination and for other possible future uses.
- SIF: Signaling information field, made up of a routing label to identify the proper signaling route and the particular transaction using originating and destination signaling point identifiers, as shown in Figure 4.26.
- Check bits: 16 bit CRC to verify the integrity of the message.

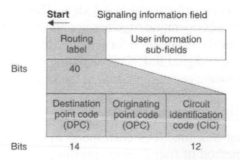

Figure 4.26
Signaling information field

Signaling network functions (layer 3)

The signaling network functions correspond to the transport layer of the OSI model which are independent of the operation of the individual signaling links. The transport functions fall into two major categories: signaling message handling functions which direct the message to the proper signaling link or user part; and signaling network management functions which control the current message routing and the configuration of the signaling network facilities.

Signal connection control part (layer 4)

The signaling connection control part (SCCP) provides additional functions to the message transfer part (MTP) to support both connection-oriented services and connectionless services between switches. This involves peer-to-peer communications between the two SCCP layers in order to setup and release the logical signaling connections, and transfer the data with or without a logical signaling connection.

Telephone user part (layer 4)

The telephone user part (TUP) defines the necessary telephone signaling functions for international and national telephone call control signaling. This information is carried in the user information sub-fields of the SIF, as shown in Figure 4.27. These fields are utilized as follows:

- *Heading codes*: Code H0 identifying as the particular message group, for example forward address messages have code 0001, forward setup 0010, backward setup 0100 etc. Code H1 contains a signal node or in the case of complex message identifies their format.
- *Calling party category*: Identifies the type of calling subscriber, data calls, test calls, and the language of the assistance operator.
- *Message indicators*: 12 bits defining various aspects of the call, such as the type of address, the presence of a satellite circuit in the connection, the presence of the echo suppressors, etc.
- *Number of address signals*: How many address signals are being sent.
- *Address signals*: Each address signal digit is sent as 4-bits, with the most significant sent first.

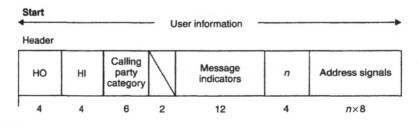

n = Number of address signals

Figure 4.27
User information field

Signalling network functions (layer 3)

5

Private switched telephone services

Objectives

When you have completed study of this chapter you should be able to:

- Explain the basic functions and components of a PBX
- Explain the functions of Centrex and the difference between Centrex and a PBX
- Explain the difference between PBXs and key systems
- Explain the functions, components, architecture, configuration and features of CTI, with particular reference to TAPI, JTAPI and TSAPI as well as first and third party call control
- Explain the following concepts:

    - Direct inward dialing
    - Direct outward dialing
    - Call accounting
    - Voice mail
    - Automated attendance
    - Unified messaging
    - Automatic call distribution
    - Call centers
    - Hospitality services.

5.1   Private branch exchange

5.1.1   Functions

A private branch exchange (PBX) is a private telephone system used within an enterprise. Users of the system share a number of outside lines for making telephone calls external to the PBX, and make internal calls via the PBX. Most medium to large companies use a PBX since it is less expensive than connecting an external telephone line to every telephone in the organization. It is also easier to make internal calls within a PBX since it can be done with typically three or four digits.

The basic functions of a PBX are very similar to those of the Central Office switch since the PBX evolved from operator controlled switchboards that were used on public telephone networks. A PBX, through control signaling, performs three basic functions:

1. In response to a call request, it establishes end-to-end connections among subscribers on the network, or to remote subscribers off the network, through intermediate nodes which might consist of other PBXs or Central Office switches on the PSTN. A circuit switched technology is used, hence the connected path is dedicated to the user for the duration of the call.
2. It supervises the circuit and detects signaling such as call request, answer, busy, and disconnect.
3. After completion of the call, it disconnects the circuit so that other users can access the resources.

Normally 1 trunk line is allowed for every 8–10 users. Depending on the call patterns of the organization, some of these trunks are then allocated as incoming lines, and the remainder as outgoing lines.

5.1.2   History

The first simple operator controlled switchboards were installed in 1878 by the Bell Company and served 21 subscribers. Since then, several generations of PBXs have emerged. These can generally be classified as one of five different 'generations'.

1. 'First generation' PBXs consisted of operator control patch panels.
2. 'Second generation' PBXs were based on step-by-step (Strowger) and crossbar switches and featured automatic dialing as well as space division multiplexing (SDM).
3. 'Third generation' PBXs included the features of the second generation PBXs but, instead of electro-mechanical control, they use semiconductor switches under stored program control.
4. 'Fourth generation' PBXs are computer-based and include functions such as automatic call distribution and voicemail. They also use time division multiplexing, which permit the integration of voice and data on T-1 and ISDN services.
5. 'Fifth generation' PBXs add LAN as well as Internet connectivity (i.e. support for the Internet protocol, IP) plus support for WAN technologies such as ATM. They also support management via the simple network management protocol (SNMP).

5.1.3   Features

PBXs typically embody the following features, many of which can be invoked from the telephone keypad:

- *Speed dialing*: This enables the user to make a call by using an abbreviated number.
- *Message waiting*: This enables the user to signal to an unattended phone that a call has been placed to it.
- *Last number redial*: This enables the user to repeat a previously dialed number by means of a single button.
- *Camp-on*: This enables the user to wait for a busy line to become idle. When this happens, a ring signal notifies both parties (caller and callee) that a connection has been made.

- *Call waiting*: By means of a special tone, a (busy) user is informed that another call has come through.
- *Call hold*: This enables the user to put the first party on hold so that an incoming call can be answered.
- *Call forwarding*: This enables a station to forward incoming calls to another station in a situation where it is busy or unattended.
- *Add-on conference*: This enables the user to make another connection while already having a call in progress.
- *Call pickup*: Enables incoming calls to unattended stations to be picked up by another station on the same trunk group.

5.1.4   PBX components

Apart from line and trunk interfaces, a typical PBX consists of three major elements namely a programmable processor, a memory (which contains both a volatile i.e. Read/Write or 'Random Access' memory and a non-volatile i.e. Read-Only memory) and a switch matrix. The switch matrix, under control of the processor, accomplishes the switching through either space division switching or time division switching.

Advanced PBXs such as the MITEL SX-2000 are designed to operate across large sites spread over a geographic area of up to several kilometers across. To accomplish this, the control unit is separated from the individual peripheral switch units by means of a high-speed fiber-optic link as shown in Figure 5.1. Several switch units (in this case to 11) can be distributed across the site and in this way up to 3000 telephones can be connected across a distributed site.

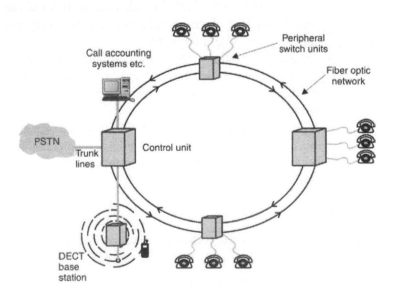

Figure 5.1
PBX configuration

5.1.5   PC-PBXs

The term PC-PBX is used to describe a telephone system based on PC technology. These systems are often referred to as communication servers because of their capabilities to provide more features and services than those found in PBXs. A typical PC-PBX provides services such as ACD, auto-attendant, voice mail, unified messaging, fax, call routing, and networking. These features will be described elsewhere in this chapter.

5.1.6   IP-PBXs

IP-PBX is a relatively new development in PBXs. They can transport intra-office voice communications over an Ethernet LANs and global voice communications over the PSTN or an IP network. Digital telephone handsets can link directly to the Ethernet LAN via 10BaseT interfaces without any physical connection to a desktop computer. Phone features can be configured using web browsers. In addition to IP networks, calls can be transported via T-1, ISDN or traditional analog lines.

Some vendors, such as Lucent technologies and Nortel, offer IP interfaces that can be retrofitted to their conventional PBX systems.

PBXs can also be interconnected over IP networks. Vendors such as PSINet, for example, offer a service whereby PBXs internal to the enterprise can be interconnected by means of the corporate Intranet.

The convergence of conventional voice switching systems and Internet technology, resulting in so-called 'converged networks' will be addressed in more detail in Chapter 9.

5.2   Centrex

Centrex (Central Office Exchange Service) is a type of PBX service where the switching occurs at a central telephone exchange (Central Office) instead of the user's premises. The telephone company owns and manages all the communications equipment, and then leases the handsets to the users. Centrex users have direct inward dialing (DID) facilities as well as station identification on outgoing calls. Each station appears as a unique line to the Central Office in the same way as a residential telephone does. Centrex users dial a four or five digit number without a prefix in order to call internal extensions, and a prefix (9) to get outside numbers.

Centrex can save money on operators, administration costs and space because of the following reasons:

- There is no switching equipment on the user's premises. All the main Centrex switching equipment is housed at the Central Office.
- It provides direct inward dialing (DID). Centrex incoming calls can go directly to the users without being answered by an operator.
- It provides direct outward dialing (DOD). On site users can dial outward calls without having an operator to place the calls for them.
- It supports automatic identification of outward dialed calls (AIOD). The telephone company bill identifies the individual telephone extensions from which outward calls were made.

There are several options available on Centrex, including the following:

- Data communications (networking)
- ISDN
- Voicemail
- e-Mail
- Message center support
- Modem pooling.

In the United States Centrex is primarily marketed at the companies with fewer than 100 lines. Apart from the obvious cost reductions mentioned before, the telephone companies promoting Centrex are quick to point out the following additional advantages.

- The supplier (telephone company) will not go out of business, with the result that support will always be available.
- Any growth in the number of extension lines is handled by the Central Office.
- Centrex is compatible with technologies such as ISDN and T-1.
- Redundancy is built into the Central Office switches, resulting in a reliable service.

Because of the Centrex networking ability, a business can set up a number of locations across a city that have a single published phone number and centralized operator or attendant service.

Users have a choice of various Centrex telephones:

- Standard '2500' sets, as used for homes. A limitation with these sets is that they only hold one or two lines
- Central Office powered phones with features provided by the Central Office
- Proprietary telephones with features provided by on-site key service units
- ISDN phones, which also get their features from the Central Office
- Specialized telephone systems for the express purpose of working with Centrex phone lines are provided by companies such as ComDial, Siemens, Nortel and Lucent.

A typical Centrex installation in a medium to large business could include:

- PC based attendant workstations
- Centrex answering consoles
- A message desk
- Call accounting system
- Administration systems.

T-1 is often used to transport Centrex service from the Central Office to the customer. Since T-1 can carry 24 voice data or video conversation over fiber optics, copper, microwave, or infra red facilities, a fiber cable with 8 fibers can carry four T-1's and serve 96 Centrex lines. In contrast, the conventional approach would need 96 pairs of copper wire to a user's premises.

5.3   Key systems

Key systems are functional replacements for PBXs in small enterprises with typically fewer than 70 users per site. Although PBXs and key systems are functionally equivalent from the user's point of view, they operate internally on slightly different principles. Modern systems are quite often key system/PBX hybrids.

A major difference between PBXs and key systems is the connection between the Central Office and the key system. PBXs are 'ground start' which means that a trunk (outside path) is seized by the PBX or by the Central Office before a call is sent between the two locations. On a PBX system, users have to dial a number (typically 9) in order to 'get a line'. A key system, on the other hand, operates on a 'loop start' in very much the same way as an ordinary analog home telephone. Pressing an outside line button on a key system signals to the Central Office that the user wants to make or receive a call.

A second difference is the way the dial tone is generated. On a PBX, the dial tone is generated by the PBX itself. On a key system, the dial tone is obtained from the Central Office.

Key systems typically vary in size from 3 Central Office (trunk) lines/8 users to 80 lines/200 users. Functions vary from basic voice processing functions such as answering machine emulation and customized greetings for each incoming line, to Web communications servers, e-mail and voicemail mailboxes, domain name management and e-commerce payment processing, to name but a few. Some key systems also have optional Ethernet, ISDN-BR/PRI and xDSL facilities built in.

5.3.1 Digital enhanced cordless telephone

Digital enhanced cordless telephone (DECT) is a digital wireless technology that originated in Europe. It is now deployed worldwide in applications such as cordless phones and wireless telephone links to homes. DECT is frequently used as an addition to a conventional PBX system in order to allow some users the ability to carry their handsets around throughout the building.

The DECT is in some ways related to GSM (global system for mobile) but is intended simply as radio access technology rather than a system architecture. It is capable of internetworking with many other systems such as the PSTN, ISDN and GSM.

As a wireless technology, DECT will be discussed on more detail in Chapter 10.

5.4 Computer telephony integration

Computer telephony integration (CTI) allows a computer and a telephone to work together so that, for example, a user can, by using a PC in conjunction with a telephone:

- See caller information 'pop-up' on the PC screen
- Set up conference calls
- Manage faxes
- Manage voice mail.

5.4.1 CTI functions

CTI was developed to use computer intelligence to manage telephone calls. The CTI functions can be classified into three categories namely call control, media processing and customer management.

These three functions will now be discussed individually.

Call control

Call control functions include:

- Call set-up and release related services such as dialing
- Call routing related services
- Network interfacing services such as tone detecting and tone generation
- Automatic call distribution (ACD) whereby incoming calls are managed in various ways including holding them in queues and handing them out to available agents.

Media processing

Media processing includes:

- Voice/fax processing such as voice recording and fax sending
- DTMF (dual-tone multi-frequency) digit processing, text to speech synthesis and speech recognition
- Call-logging such as online recording and call accounting.

Customer data management

Customer data management involves automatic number identification (ANI) in order to provide the system with the telephone number of an incoming call. This is used for retrieving the information pertaining to the caller from the data base (data base matching). Information contained in the database may include phone books, schedules and billing records.

5.4.2   CTI components

The CTI architecture consists of three components. They are:

1. The switch-to-host interface which provides connection between the telephone switch and the controlling host (the CTI server)
2. The application programing interface (API) which allows a software developer to create new functions and services for the system
3. The CTI resource architecture which manages the telephony and computing resources (building blocks) such as speech recognition and fax boards.

These building blocks will now be discussed in more detail.

The switch-to-host interface

The switch-to-host interface provides the connection between the switch and the host computer. The switch is connected to the host via a CTI link and enables programmers to control the switch via standard programing interfaces such as TAPI and TSAPI which will be discussed later.

There are two technologies that can be used for the CTI link interface namely computer supported telecommunication applications (CSTA) and switch-to-computer application interface (SCAI). The first technology (CSTA) was initiated by the European Computer Manufacturers Association (ECMA) and has recently become the de facto industry standard. CSTA is the base on which TSAPI is defined.

The application programing interface

The second component of the CTI architecture is the application programing interface (API). This is an interface that enables software developers to create new applications. Three standards will be discussed namely TAPI, TSAPI and JTAPI.

TAPI

Microsoft's telephony application programing interface (TAPI) enables Windows programmers to develop telephony applications in software. The current version is TAPI 3.0. The TAPI 3.0 component object model allows programmers to write TAPI applications in, for example, JAVA, Visual Basic or C++. TAPI 3.0 also supports IP telephony (VOIP) that complies with the ITU-T H.323 standard.

TAPI enables an application to set up calls, terminate calls, monitor call progress, detect calling line identification, hold calls, transfer calls, conference, park, and pick-up. It can re-direct calls, forward calls, answer and route incoming calls. It provides access to various telephone services such as the POTS, ISDN, digital network services that support data communications, and other services such as Centrex.

TSAPI

While TAPI defines the connection between a single phone and a PC, the Novell NetWare telephone services application programing interface (TSAPI) defines the connection between a NetWare file server and a PBX. The physical link is an ISDN BRI card in the server that connects with the PBX.

TSAPI was developed by Novell and Lucent technologies as a NetWare-loadable module (NLM) that resides in a Novell server to support a 'call model'. It is focused primarily on a so-called third party call control and its main application is the call center environment.

JTAPI

The Java telephony API (JTAPI) provides an object oriented API for Java-based computer telephony applications. It supports both first party call control and third party call control to implement applications such as distributed call centers.

CTI resource architecture

The third component of the CTI architecture is the CTI resource architecture. Within a CTI system, hardware resources are required to implement the various functions. These resources include analog and digital trunk interface boards, voice processing boards, fax boards, speech synthesis boards and speech recognition boards. A CTI resource architecture is essentially an open system that supports interoperability among various CTI resources made by different suppliers or vendors. The standards for the CTI resource architecture include MVIP, SCSA and ICTF H.100/H110.

MVIP

Multi vendor integration protocol (MVIP) is a family of open standards allowing integration of PC technology with proprietary PBXs and other telephone switching systems. The original MVIP standard (MVIP-90) specifies a TDM bus with 512×64 kbps capacity and the circuit switching capability inside the computer under software control. A compatible superset of MVIP-90, namely H/MVIP (High Capacity MVIP) extends the system to handle up to 3027×64 kbps.

SCSA

Signal computing system architecture (SCSA) is another open architecture for integrating computer telephony resources. The SCSA software model is called the telephony application objects (TAO) framework and identifies a set of interfaces and services for manipulating the media of a call (play, record, etc.). The hardware model defines the SCbus, which supports 1024 and 2048 TDM switching time slots for ISA and VME bus systems.

ICTF H.100/H.110

The Enterprise Computer Telephony Forum specifies a high capacity bus H.100/H.110 bus to provide compatibility for existing products. The bus features:

- 4096 time slots at 8 MHz (vs 2048 for SCbus and 512 for MVIP-90)
- Redundancy through dual clock and frame synchronization lines so that if one clock line is lost all devices can synchronize with the other clock
- Interoperability: 16 of the 32 data lines can be selectively downgraded to 2 MHz for MVIP-90 or 4 MHz for SCbus.

H.100 is the specification for ISA bus while H.110 is the specification for the CompactPCI (cPCI) bus. Additional features of the CT bus on cPCI are 32 data streams at 8.192 Mbps, over 260 Mbps bandwidth, redundant clocks, and maximum 20 slots.

5.4.3   CTI configurations

This section describes the two possible CTI configurations, namely first party call control and third party call control.

First party call control

In the first party configuration a call is controlled by the call parties. First party or phone oriented call control involves a direct interface between the user's PC and the telephone switch by intercepting line signaling between the telephone and the switch as shown in Figure 5.2. Because of this approach, it limits itself to conventional signaling conditions used in PSTN telephones. The telephone-to-computer interface is usually a phone line connection. This connection can be via RJ-11 phone jack, RS232 or universal serial bus (USB). This type of call control is ideal for small offices and is independent of the telephone type. It can therefore interface analog, ISDN or digital telephone ports in a switch.

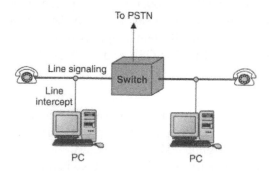

Figure 5.2
First party call configuration

Third party call control

In the third party configuration, a call is initiated from a device that does not participate in the call. The third party or switch oriented call control system controls the switch through a separate X.25 connection using a CTI link protocol such as CSTA as shown in Figure 5.3. The call control can be performed by an API such as TSAPI. The TSAPI enabled application program is viewed as an external agent (third party) in a conversation between the calling and the called parties.

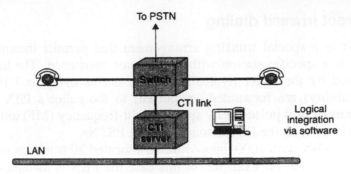

Figure 5.3
Third party call configuration

Third party control is more ideal for large sites because it is not necessary to buy hardware and software for each PC or desktop.

5.4.4 CTI features

The following are typical features associated with a CTI system.

Integrated voice recognition

Integrated voice recognition (IVR) is typically used in transaction based applications such as ticket reservation systems and account information query systems. The user accesses the system through the telephone set. The system interprets the user's instructions to the computer and the computed results are converted back to voice.

Screen Pop

Screen Pop (screen synchronization) delivers a call to an operator together with a customer's account screen. Screen Pop is enabled by identifying the caller from the calling telephone number (caller ID) or from a customer dialed PIN number. When the call is transferred to another operator, the original screen is automatically transferred as well. This concept is referred to as call/data association.

Unified messaging system

The unified messaging system (UMS) server integrates an e-mail server with voice and fax mail servers to provide a unified mail box that can receive messages of various kinds.
This concept will be dealt with in more detail in a separate section.

Call center

A CTI-based call center supports automatic call distribution (ACD) with features such as call distribution, call queuing, on hold announcement and other functions such as call transfer and conferencing. This concept, again, will be discussed in more detail in a separate section.

5.5 Other services

The following services or functions could be available on PBXs and key systems. It depends on the manufacturer and model, and obviously has a price implication. Some of the functions are also available on Centrex.

5.5.1   Direct inward dialing

This is a special trunking arrangement that permits incoming calls from the PSTN to reach a specific station without operator assistance. The last few digits in the number dialed by the caller (typically 3 or 4, but as many as 7 if the Central Office has the capability) are forwarded (outpulsed) to the callee's PBX on a special DID trunk by means of dial pulses or by special multi-frequency (MF) tones. These tones are not to be confused with the DTMF tones used on PSTNs.

A PABX with 1000 lines could be allocated 20 trunks as well as the numbers 712–1000 to 712–1999, for example. In this case the PBX is identified by 7121 and the extension lines by 000–999. If a caller dials any number in this range, e.g. 712–1555, the Central Office tries to find an available trunk (1 of the 20) and forwards the call to the PBX. The 20 trunks are equivalent, and said to be in a hunt group or a rotary group. The PBX uses the 555 sequence to identify the designated end user.

The DID trunks mentioned above could either be terminated on a PBX, as described above, or a fax server that forwards the fax, based on the number dialed, to the recipient's PC for subsequent retrieval.

DID addresses the following customer needs:

- It reduces dependence on a full-time operator
- It reduces the workload of the operator
- It gives departments and individuals the status of an individual line and in the process also creates a personal office atmosphere.

DID does, however, have a few limitations which should be kept in mind:

- Although the callee might be available, all the trunks may be busy at a given moment, in which case the caller will get a fast busy tone.
- When the callee's station is disconnected, its associated 7-digit number cannot be intercepted by the Central Office. The call is therefore directed over the DID trunks and can either be directed at the customer's PBX operator in which the calling party is charged for the call. Alternatively, the call could be directed to the customer's PBX intercept tone or recorded announcement in which case the calling party is not charged.

DID is primarily targeted at industries such as government institutions, hospitals, hotels, utilities, financial institutions, manufacturing, education, distribution and transportation.

There are also some new DID markets emerging, including voice messaging firms and telephone answering services.

5.5.2   Direct outward dialing

Direct outward dialing enables specific phones on the PBX to compete a call to any other user on the PSTN. This process it is not handled by an operator.

5.5.3   Call accounting

Call accounting is useful for tracking employee productivity, the use (and abuse) of the telephone system by employees, billing for professional services, as well as ensuring that billed calls were actually made. Typically, call accounting software packages run on a PC (with Windows) connected to the station messaging detail reporting (SMDR) port of a phone system. If the phone system does not have an SMDR port, a hardware box such as Sandman's 'Whozz Calling' hardware can be placed on the trunk lines in front of the system as shown in Figure 5.4.

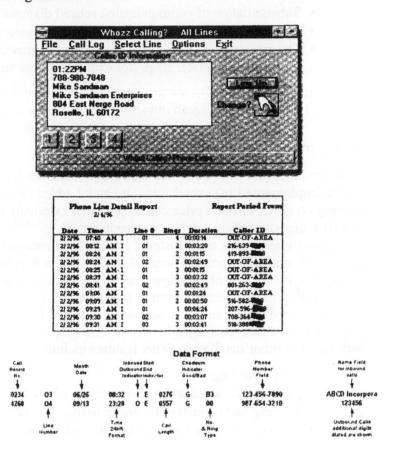

Figure 5.4
Examples of call accounting system displays (Courtesy Mike Sandman)

Typical data supplied by the system include:

- Receipts for hotel room telephone usage
- Inbound caller ID
- DTMF digits dialed on inbound calls
- Complete digits dialed for outbound calls
- Time, date and duration for incoming and outgoing calls
- Real time data for unhook, offhook and ring signals which can be used for diagnostics.

Through the appropriate software, it is possible to have pop-ups screens with full caller details on PC screens. With this information, the following can be achieved:

- The scheduling of personnel and telephone usage based on the time of day most calls are made or received.
- Checking whether inside sales staff is maintaining sufficient call volume.
- An analysis of average call durations to assess the needs of customer service personnel.
- Identification of personnel who are using or abusing the phone system.

- The allocation of telephone expenses to individuals.
- The elimination of rarely used circuits or trunks.
- The identification of failed circuits or trunks.
- Substantiation of communication related disputes.
- The identification of inbound and outbound security breaches.
- The provision of names and numbers of parties calling in after hours to enable call returns at a later stage (for example during business hours).

Some call accounting systems are specifically geared for the hospitality industry (for example hotels) and are geared towards specific use for accurate billing of guests.

5.5.4 Voice mail

Since its inception in 1980, voice mail has become an effective communications tool. It can be implemented in several ways. It can be added on to a PBX by means of software running on a PC, or the service can be provided externally either by a telephone company or a firm that specializes in voice messaging.

Business class voice mail systems for small to medium sized organizations typically come with multiple ports (between 2 and 8) for interfacing with the telephone system, storage for up to 50 h of messages and an unlimited number of password protected mail boxes. Users can access the system to customize their personal greetings and use dial codes to record, check and send outgoing messages. Other dial codes enable users to play, skip, save or delete messages locally or remotely. Copies of received messages can be forwarded to other mailboxes. Other features include:

- Call screening
- Pager notification when messages have been left
- Playing messages faster, slower, louder or softer
- Messaging options such as private, urgent, etc.

Modern voice mail systems are typically administered via Windows based set-up screens, running on a laptop or desktop computer.

5.5.5 Automated attendant

Automated attendants, also referred to as 'nests', are voice mail systems without operators that provide routing of calls through simple menus. For example, a menu might say 'for customer service press 1, for sales press 2 . . .'. Since routine questions can be answered by the automated system, a department's overall call handling ability is increased and employees are free to answer more detailed questions.

5.5.6 Unified messaging

Unified messaging is the integration of several different communication media such as voice, fax and e-mail messages through a single interface, be it a normal telephone, wireless phone or PC as shown in Figure 5.5. In the words of the Unified Communication Consortium, it is 'the ability to create and respond to multimedia messages with fidelity to the originator from either a telephone or PC, especially across different vendor platforms. Additionally, personal call control permits real time control of incoming calls and call rebound with message processing.'

There is a proliferation of devices people must use to receive all their messages. These include desk phones, cellular phones, fax machines, alphanumeric pagers and e-mail systems. A unified messaging system consolidates this by depositing each subscriber's e-mail, fax and voice messages into a universal messaging inbox so that the subscriber can find his messages in a single place through a single interface such as a web browser or desktop application such as Lotus Notes.

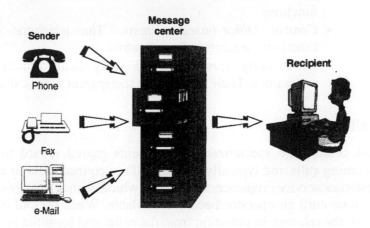

Figure 5.5
Unified messaging

Unified messaging systems notify subscribers whenever a new message arrives, whatever the medium might be. Notification methods include e-mail notification, pager, a stutter dial tone, or a message waiting indicator light.

With a web browser such as Netscape Navigator or Microsoft Internet Explorer, subscribers can view fax and e-mail messages on their screens and listen to voice messages over headphones or speakers.

5.5.7 Automatic call distribution

Automatic call distributions (ACDs) are typically used by companies that have call centers with high incoming call rates and that have several agents whose sole responsibility is handling incoming calls. ACDs are typically used in the following customer services, credit authorization and reservations environments.

ACDs greet callers with pre-recorded messages, provide dial options and then route calls to the appropriate individuals. They can provide music to callers who are on hold and can make announcements regarding queue status. Some ACDs can also handle incoming calls from the Internet and serve up to 50 agents in a call center environment.

ACD systems consist of incoming lines, agent positions and the switch itself. Normally toll free –800 lines are connected to ACD systems but any type of line, including ISDN, can be used.

The following types of services are available:

- Stand-alone systems are used where the service center is separate from the rest of the business and ACD systems are not integrated with the telephone system.
- Integrated systems comprise a PBX key telephone system with additional ACD software providing the call allocation and supervision within the telephone system itself.
- PC based systems comprise software on a multi media equipped PC connected to the PBX.

- Automatic call sequencers are independent devices that perform the same type of allocation as an ACD but without complex time and load calculations. These systems have to rely on the PBX for routing calls because they do not have their own switch matrix.
- *Centrex-based systems*:   In this case the telephone company provides ACD functions.
- *Central Office-based systems*: The telephone company provides ACD functions, separately from Centrex.
- *Third party servers*: Third party firms provide ACD service to other companies. Their operation is transparent to the caller.

5.5.8   Call centers

Call centers are specialized environments geared toward managing large volumes of incoming calls and typically has an ACD to connect calls to an order taker, help desk or customer services representative. Calls which cannot be answered immediately are put in a queue until an operator becomes available. When the call is answered, the agent takes down the relevant information from the caller and feeds the information into the computer database. A typical call center includes the following:

- Multi-button telephones for the keyboard entry of customer information. These telephones are equipped with a headset for hands-free operation.
- Workstations that enable agents to enter custom information on their screens.
- A host computer and database for storing customer information.
- The switching system which is most often a stand-alone ACD or ACD/PBX system.
- Telephone lines and services, such as 56K, T-1 or Fractional T-1, and ISDN. Services typically consist of –800 toll free numbers or –900 service providing numbers.
- Internet connection. A recent development in e-commerce is the integration of the Internet with the call centers, enabling clients to converse with call center operators via their browsers at a click of the mouse.

5.5.9   Hospitality systems

A hospitality system (such as the Alcatel Office Guest) is purpose designed for the hospitality industry such as hotels, but is equally applicable to clinics, retirement homes and student residences. Such a system typically offers the following features:

- *Front desk services*:   These include direct guest calling by name or by room, a global overview of all rooms (availability/status), checkout at guest departure, etc.
- *Guest services*:   These include DID to the room and direct access to hotel services (front desk, bar, restaurant, etc.) from rooms via predefined keys
- Do not disturb activation from front desk or room
- Mail box services for rooms
- Wake up services from the front desk or from rooms
- Mobility for special guests or hotel staff through DECT cordless phones
- Prepayment services
- *Cost control*:   This includes telephone calls using personal PIN numbers
- Barring (e.g. no direct calls, local, national or international, etc.).

5.5.10   Miscellaneous other options

Class-of service restrictions

This function allows control over certain services or shared resources. Access to long distance services, for example, can be restricted by country code, area code or exchange. Access to the modem pool for transmission over analog lines can be similarly controlled.

Hunting

This is a capability that routes call to an alternate station when the called station is busy.

Music on hold

This indicates to callers that the connection is active while the call waits in queue for the next available operator.

Power-fail transfer

This capability is used in conjunction with an uninterruptable power supply (UPS) and allows the continuation of communication to the external network during a power failure.

System redundancy

This involves the sharing of the switching load amongst more than one processor. In case of a processor failure, other processors will share the increased load.

6

Public transport network technologies

Objectives

When you have completed study of this chapter you should be able to:

- List the functions of the various components of a modem
- Discuss the effects of line distortion on modem performance
- Discuss suitable applications for dial-up modems
- Describe the advantages in using conditioned lines for modems
- Explain how subrate multiplexing operates on T-1 systems
- Explain how the process of drop-and-inset multiplexing is achieved
- Describe the functions of the ISDN D-Channel
- Explain the use of virtual circuits in frame relay systems
- Discuss applications for switched multi-megabit data services (SMDS)
- Discuss cell switching in asynchronous transfer mode (ATM)
- List the speeds of the synchronous data hierarchy (SDH/SONET).

6.1   Overview

In this chapter we will now consider the different communication approaches used for the provision of services in the public telecommunications network. Fundamentally circuits can be provided in four different categories: either analog or digital and each of these can be either switched or leased. These will be covered in the following sections and the various alternatives in each category will be discussed. First we will consider the switched analog network and then look at the use of dedicated or leased analog circuits. Next the digital services will be discussed beginning with the various types of switched digital services and concluding with various alternatives for dedicated digital services, primarily directed at the large users. In each section the method of operation will be explained, the circuit characteristics described and the particular advantages the method offers will be considered. This is intended to enable the reader to make an informed choice about the appropriateness of that particular approach for their unique situation.

6.2   Switched analog services

The public switched telephone network (PSTN), as discussed in Chapter 4, provides switched analog services. Here the customers are connected to the telephone exchange by a twisted-pair copper cable with about 3 kHz of bandwidth, typically 300–3400 Hz. Depending on the type of telephone exchange, the subscriber's analog signals may be converted to digital transmission to go across the network and then back to analog again, or in older exchanges they may remain as analog throughout. In either network the connection is controlled by the subscriber's telephone instrument and the transmission to and from the local exchanges uses an analog circuit. This is illustrated in Figure 6.1. While these networks were developed for the transmission of voice traffic, they can also be used for transmission of digital data.

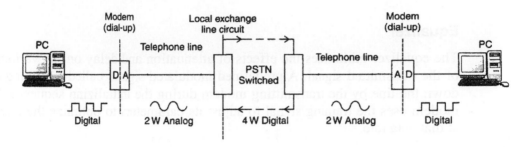

Figure 6.1
Switched analog connection

If the subscriber wishes to convey digital data across of this circuit, then a modem (modulator–demodulator) is required in order to convert the digital data into analog signals which will pass through the limited bandwidth of the telephone channel.

6.2.1   Modem components

A block diagram of a modem is shown in Figure 6.2. The functions of the various components are as follows:

Transmit buffer

The transmitted data is stored in the transmit buffer of the modem.

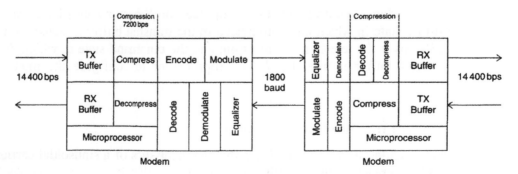

Figure 6.2
Modem components

Compression

The modem's microprocessor uses a compression algorithm to compress the transmitted data in the transmit buffer. The amount of compression achieved depends on the compression algorithm used for the applicable modem standard and is typically about 2:1. In this case a 14 400 bps data stream would then be effectively transmitted at 7200 bps.

Data encoder/modulator

The data encoder/modulator takes the serial bit stream and using multilevel encoding suitably modulates analog tone(s) for transmission over the telephone line. For example with QAM each electrical transition would represent 4 bits, so the above signal would be transmitted at about 1800 baud.

Equalizer

The equalizer minimizes the effects of attenuation and delay on the various components of the transmitted signal. A predefined modulated signal, called a training signal, is sent down the line by the transmitting modem during the initializing sequence. The receiving modem uses the training signal to adjust its parameters to optimize the line for reception at that data rate.

Demodulator/decoder

The demodulator retrieves the bit stream from the analog signal and decodes it to return the original data.

Decompression

The microprocessor in the receiving modem uses the same algorithm to decompress the received data and places this in the received buffer.

Received buffer

The data in the received buffer is sent out at the original data rate of 14 400 bps.

Microprocessor

Each modem is controlled by a microprocessor. Smart modems have two main states: the on-line state in which any data placed in the transmit buffer is transferred to the received buffer over the telephone connection; and the command state in which data placed in the transmit buffer is used by the microprocessor to configure the modem or undertake testing.

6.2.2   Modem modulation methods

The modulation process modifies the characteristics of a sinusoidal carrier signal, which can be represented by the following equation:

$$V(t) = A\sin(2\pi ft + \varphi)$$

Where

V(t) = the instantaneous voltage at a time t
A = amplitude of signal
f = frequency
φ = phase angle.

The modulation techniques used in modems include:

- *Amplitude modulation or amplitude shift keying (ASK)* in which the binary values are represented by two different amplitudes of tone. ASK is not used alone in modern modems.
- *Frequency modulation or frequency shift keying (FSK)* is commonly used for modems at data rates less than 1200 bps, where different frequencies are allocated to the logic one and the logic zero in the data message. Full-duplex modems use four frequencies, one pair for each direction of transmission. FSK modems are very reliable but need to transmit one tone for each data bit. The rate at which this can be done is called the baud rate and is limited by the bandwidth of the channel.
- *Phase modulation or phase shift keying (PSK)* uses changes in phase angle to represent the different data bits. For example QPSK uses four phase angles, and so one electrical transition represents two bits of information. In this way the data rate is twice the baud rate, so a 2400 bps data rate is carried at 1200 baud. Phase shift keying is used in some form in all high-speed modems.
- *Quadrature amplitude modulation (QAM)* uses a combination of simultaneous phase and amplitude modulation to convey even greater data rates. QAM allows data rates of four bits per baud, so a 9600 bps data rate is carried at 2400 baud. QAM systems usually incorporate trellis coding in which only some of the possible phase/amplitude combinations are valid. This means that if the presence of noise on the line causes the received signal to differ from an accepted combination, then the receiver will choose the nearest valid point. This usually also involves forward error correction as well. These systems then have high accuracy whilst transmitting many bits for each electrical signal transition.

6.2.3   Line distortion

There are two significant causes of distortion to the signal during data communication, which are illustrated in Figure 6.3. These are:

- Attenuation distortion
- Envelope delay distortion also known as phase distortion.

Attenuation distortion

Attenuation distortion is caused by the fact that not all frequencies across the passband of a channel have the same degree of attenuation. The ITU-T Rec. G.132 recommends no more than 9 dB of attenuation distortion, relative to the reference level at 800 Hz, across the passband between 400 and 3000 Hz. The higher frequencies tend to be attenuated more easily and attenuation becomes more non-linear at the edges of the passband. The equalizer compensates with an equal and opposite effect to give the channel constant loss across the passband as shown in Figure 6.3.

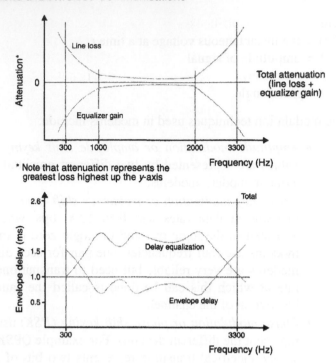

Figure 6.3
Attenuation distortion and envelope delay

Envelope delay distortion

Envelope delay distortion occurs because not all frequency components in the input signal propagate through the channel to arrive at exactly the same time. This can cause problems when the baud rate on the channel has increased to such an extent that the delayed frequency components of one symbol overlap with the following symbol causing intersymbol interference. This is the main limiting factor on the modulation rate for high-speed data transmission and it is desirable to limit delay distortion to less than the transmission time for one symbol.

6.2.4   Smart modem AT commands (Hayes compatible)

All smart modems utilize the Hayes compatible commands to control their operation. The smart modem is put into the command state by the sequence of characters: '+ + +' and then the attention command: 'AT'. This is then followed by the command codes, and hundreds of combinations have been developed to cover the many options available on modern modems. It should be noted that all smart modems utilize the same basic character set, but the more sophisticated commands are not always standardized and the user needs to carefully check the availability and syntax of particular commands. A few of the more common commands are as follows:

A   answer mode, answers incoming call immediately
Dn   dials telephone number n
T   uses tone dialing
P   uses pulse dialing
W   waits for dial tone
H0   forces modem to hang up

H1 forces modem off-hook
L controls speaker volume
M turns speaker on or off
Sn controls status register n – used for configuration
Z reset modem profile
&Tn used for testing purposes
%Cn changes data compression method.

6.2.5 Microcom networking protocol

The microcom networking protocol (MNP) was developed to enable smart modems to optimize their transmission performance by negotiating the best method of error detection, data compression and transmission speed for the data transfer over the chosen telephone circuit. There are nine MNP Classes defined in Table 6.1, which cover the transmission alternatives. When two smart modems begin a connection one of the modems sends a link request advising its MNP class and the other replies with its details. Both modems select the highest MNP class that both can support, then one modem sends a training signal to the other to optimize its equalizer setting for the chosen line and data rate and the procedure is repeated for the opposite direction. Once both equalizers have been adjusted, the data transfer is checked and if transmission quality is acceptable the message transfer begins. If errors are encountered the modems adopt a fall-back strategy: selecting a lower data rate, retraining the link and verifying the data transfer. The highest quality modems also will operate a fall forward strategy by pausing during data transmission to renegotiate a higher data rate, if a long sequence of frames is received without errors once the modems have fallen back to a lower data rate.

MNP Class	Async/ Synchronous	Half or Full Duplex	Efficiency (%)	Description
1	Asynchronous	Half	70	Byte oriented protocol
2	Asynchronous	Full	84	Byte oriented protocol
3	Synchronous	Full	108	Bit oriented protocol Communications between (PC) terminal and modem is still asynchronous
4	Synchronous	Full	120	Adaptive Packet Assembly (large data packets used if possible). Data phase optimization (elimination of protocol administrative overheads)
5	Synchronous	Full	200	Data compression ratio of 1.3 to 2.0
6	Synchronous	Full	–	9600 bps V. 29 modulation universal link negotiation allows modems to locate the highest operating speed and use statistical multiplexing
7	Synchronous	Full	–	Huffman encoding (enhanced data compression) reduces data by 42%

(Continued)

MNP Class	Async/ Synchronous	Half or Full Duplex	Efficiency (%)	Description
8	Synchronous	Full	–	CCITT V.29 fast Train Modem technology added to class 7
9	Synchronous	Half-duplex emulates full duplex	–	CCITT V.32 modulation + Class 7 Enhanced data compression. Selective retransmission in which error packets are retransmitted

Table 6.1
MNP protocol classes

6.2.6 Modem selection considerations

Some of the more important modem features that need to be considered when selecting a modem are as follows:

- *Automatic smart features*:   Modems normally use the Hayes AT command set, which automates most modem features. If compatibility problems arise some of the features can be manually overridden.
- *Data rate*:   This is normally one of the first features chosen. It is important to distinguish between the effective data rate and the baud rate. Note that more reliable performance is obtained at lower data rates because of the correspondingly lower baud rates.
- *Asynchronous/synchronous*:   Higher speed data rates are generally achieved using synchronous transmission, however if the line is noisy then asynchronous transmission can achieve greater throughput.
- *Full-duplex/half-duplex*:   Full-duplex transmission is normally more efficient than half-duplex where the line turnaround time reduces efficiency.
- *Modem standards*:   Most modems support a range of ITU standards such as V.22 bis, which supports 1200 and 2400 bps transmission, V.32 bis which supports 4800/9600 and 14 400 bps transmission, V.34 which supports 28 800 bps and V.90 which supports downloads at 56 kbps. In addition many of the Bell series of modem standards are also supported.
- *Data compression*:   The data compression algorithm is selected by the MNP class, and may be overridden by modem command.
- *Error correction/detection*:   The error detection and correction mechanisms are implicit in the MNP class and the chosen modem standard. These are normally auto-negotiated but may be overridden if problems are occurring.
- *Mounting*:   Modems are available in various forms of mounting such as PC internal, external or rack mounted.
- *Power supplies*:   Modems can have individual power supplies or can be supplied from a common rack supply.
- *Testing*:   Modems have varying degrees of self testing available, and local and remote loopback tests are normally available using AT&Tn commands for localizing faults.

6.2.7 Applications of dial-up modems

The principal applications for dial-up services are to provide casual access to the Internet and also to allow remote access into corporate computer systems. Dial-up access enables connections to be made quickly, on demand, from any telephone and this includes mobile telephone service. The cost effectiveness of this method of access depends on the telephone network charging regime. If calls are charged on the basis of a fixed connection charge plus a call duration time component, then depending on the relative size of these charges, it may be economic to make a number of short calls or conversely to make one long call and hold the circuit between transactions.

Today, most remote access users are employees working away from the office or those authorized to telecommute from their homes, who need remote access to the organization's network and its resources. This is done using a dial-up modem on the remote user's PC or laptop. The office uses a modem or pool of modems, connected to a remote access server on the organization's network. The user simply dials into the network equipment and the connection is made, after a security check for the authenticity of the user.

Dial-up remote access can also be used for the connection of point-of-sale (POS) transaction machines, such as credit card readers, cash registers, and ATM (Automated teller machine) equipment. The use of on-demand dial-up circuits and modems to connect ATM machines, banks, and credit-approving networks may be cost-effective for low volume users. Whenever a credit card check is made on a small POTS-oriented credit card reader, an on-demand connection to a credit-approving network can be made. A similar transaction-based network can be used by various state-run lotteries. Whenever the vendors are involved in small numbers of ticket sales, an on-demand POTS circuit is often used to convey the details to the central lottery computer.

6.3 Leased analog data services

Dedicated or leased analog lines are available for permanent connection across the PSTN. Most of these lines are used for data services although others are used for permanent analog connections such as for example music distribution or control circuits. In this section we will only consider the special requirements for data applications. To utilize these circuits for data, a leased line modem is used in a similar manner to dial-up circuits. The only difference between a leased line and dial-up modem is in the method of the signaling. A leased line modem does not need to connect to any plant in the local telephone exchange, as all we are leasing from the telephone company is a pair of wires between the two locations. This is illustrated in Figure 6.4.

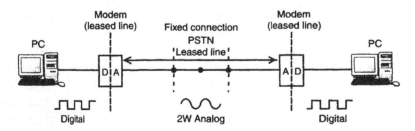

Figure 6.4
Leased line modem connection

6.3.1 Leased line modems

The leased line modem is permanently connected and does not require any signaling such as dial tone, ringing, etc. needed for signaling purposes in dial-up modems. The modems are configured using the AT commands already discussed in Section 6.2.4. Some leased line modems can also be used on dial up lines. One modem is set to answer mode, while the other is set to originate mode. This ensures that the modems will be transmitting and receiving on the correct frequencies, etc. The frequencies used in the ITU V.21 and the Bell 103/113 standards are illustrated in Table 6.2.

Specification	Originate (Hz)		Answer (Hz)	
	Mark	**Space**	**Mark**	**Space**
ITU V.21	1270	1070	2225	2025
Bell 103	980	1180	1650	1850

Table 6.2
Leased line modem frequency allocations – V.21 and Bell 103

6.3.2   Conditioned lines

The leased circuits used for data transmission can be fitted with equalizers to improve the quality of their transmission characteristics to make them more suitable for data transmission. These circuits are said to be conditioned and various grades of conditioning are available to improve the frequency response and the envelope delay distortion. The differing grades of C-type conditioning available from AT&T are illustrated in Table 6.3. C-type conditioning applies only to frequency response and delay distortion characteristics, while D-type conditioning applies higher standards with regard to various types of harmonic distortion as well.

Conditioning Type	Frequency Response Relative to 1004 Hz		Envelope Delay Distortion	
	Frequency Range	**Variation in dB**	**Frequency Range**	**Variation in microseconds**
Basic	500–2500	−2 to +8	800–2600	1750
	300–3000	−3 to +12		
C1	1004–2404	−1 to +3	1004–2404	1000
	304–2704	−2 to +6	804–2604	1750
	304–3004	−3 to +12		
C2	504–2804	−1 to +3	1004–2604	500
	304–3004	−2 to +6	604–2604	1500
			504–2804	3000
C3 (Trunks)	504–2804	−0.5 to +1.5	1004–2604	110
	304–3004	−0.8 to +3	604–2604	300
			504–2804	650
C4	504–3004	−2 to +3	1004–2604	300
	304–3204	−2 to +6	804–2804	500
			604–3004	1500
			504–3004	3000
C5	504–2804	−0.5 to +1.5	1004–2604	100
	304–3004	−1 to +3	604–2604	300
			504–2804	600

Table 6.3
AT&T C-type line conditioning examples

6.4   Digital transmission hierarchies

Digital transmission systems have the particular advantages of excellent noise performance enabling signals to be taken virtually unlimited distances without signal degradation, better error performance, faster systems, more cost-effective switching and multiplexing and better integration with data and video services, etc. The PSTN is progressively evolving towards all-digital transmission. Unfortunately, different time division multiplexing (TDM) standards have evolved for these systems. In the United States, Canada and Japan the DS-1 transmission format has evolved for multiplexing 24 voice channels together, whereas the alternative method of multiplexing 30 channels is adopted by most other countries as the E-1 format complying with the ITU-T recommendations. These two multiplexing formats will now be described.

6.4.1   DS-1 (T1) multiplexing format

The DS-1 or T1 multiplexing format is built around 24 channels containing 8-bit samples repeated 8000 times per second which yields an aggregate bit rate of 1.544 Mbps. Additional bits are incorporated in the frame to allow for synchronization and for the signaling associated with the voice channels. The DS-1 frame format uses a single bit at the start of each frame for synchronization and this bit alternates between 1 and 0 in consecutive frames. For voice services frames 6 and 12 use 7-bit speech samples and the eighth bit is used for signaling. These bits form data streams for each voice channel which are used to establish or terminate connections, etc. The voice frame format is illustrated in Figure 6.5.

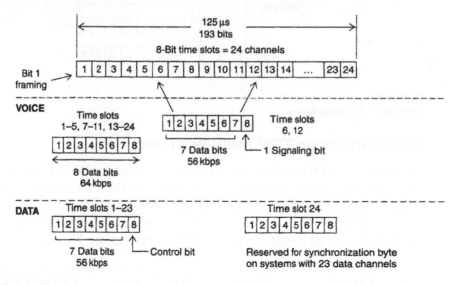

Figure 6.5
DS-1 transmission frame format

The same DS-1 frame is used for digital data service at 1.544 Mbps. Here 23 channels of data are sent per frame and the 24th channel is reserved for a special synchronizing byte to allow faster and more reliable reframing following a framing error. In each channel, seven bits are used for data and the eighth bit is used to indicate whether the channel contains user data or system control data. By using seven bits per channel repeated 8000 times per second each channel can support a data rate of 56 kbps. If the DS-1 system is used to carry voice and data then channel 24 is not required for synchronizing and all 24 channels are available. The data frame format is illustrated in Figure 6.5.

6.4.2   Fractional T1 systems

Fractional T1 systems can provide lower data rates by using a technique called subrate multiplexing. Here an additional bit is taken from each channel to indicate which subrate channel speed is being provided. This now leaves each channel with only six bits and at 8000 frames per second which results in a maximum speed of 48 kbps. This capacity can be used to multiplex five channels at 9.6 kbps or 10 channels at 4.8 kbps or 20 channels at 2.4 kbps. In this method the same channel is used for all the subchannels, with each subchannel sending its six data bits in consecutive frames, so for example one 9.6 kbps channel would send its data every fifth frame. This is illustrated in Figure 6.6.

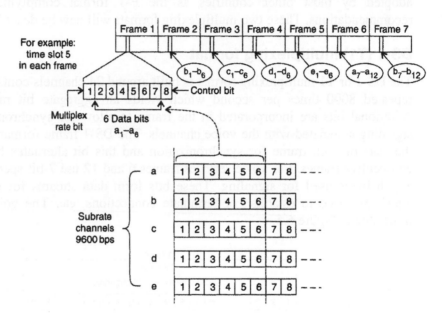

Figure 6.6
Fractional T1 multiplexing

6.4.3   North American TDM multiplex hierarchy

Higher level TDM multiplexing is used to combine DS-1 systems by interleaving bits from the various DS-1 inputs. For example, the DS-2 transmission system combines four DS-1 inputs into a 6.312 Mbps data stream byte interleaving 12 bits at a time from each stream. Additional bits are required for framing and control purposes. Table 6.4 illustrates the various circuit designations and their corresponding bit rates.

Circuit Designation	Voice Channels	Data Rate (Mbps)	Circuit Combinations		
DS-0	1	0.064	–	–	–
DS-1	24	1.544	24 DS-0	–	–
DS-1C	48	3.152	2 DS-1	–	–
DS-2	96	6.312	4 DS-1	2 DS-1C	–
DS-3	672	44.736	28 DS-1	14 DS-1C	7 DS-2
DS-4	4032	274.176	168 DS-1	84 DS-1C	42 DS-2

Table 6.4
North American TDM multiplex hierarchy

6.4.4 ITU-T E1 multiplexing format

The E1 multiplexing format adopted by the rest of the international telecommunications community uses 30 traffic channels, plus a synchronizing channel and a signaling channel, each of 64 kbps giving an aggregate bit rate of 2.048 Mbps. The E1 frame format is illustrated in Figure 6.7. Synchronization uses time slot zero and all the signaling information for the 30 channels is carried in the time slot 16. Fractional E1 systems allow bit rates of less than 64 kbps.

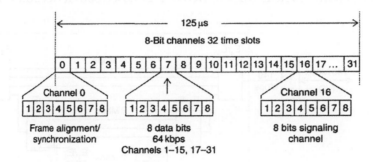

Figure 6.7
E1 multiplexing frame format

6.4.5 European TDM multiplex hierarchy

Higher level TDM multiplexing is used to combine E1 systems by interleaving bits from the various E1 inputs. For example, the E2 transmission system combines four E1 inputs into an 8.448 Mbps data stream byte interleaving. Additional bits are required for framing and control purposes. Table 6.5 illustrates the various ITU-T circuit designations and their corresponding bit rates.

Circuit Designation	Voice Channels	Data Rate (Mbps)
E1	30	2.048
E2	120	8.448
E3	480	34.368
E4	1920	139.264
E5	7680	565.148

Table 6.5
European TDM multiplex hierarchy

6.4.6 Plesiochronous digital heirarchy

The plesiochronous digital heirarchy (PDH) describes the above systems of high order multiplexing, these systems are also sometimes called asynchronous multiplexing. The term 'plesiochronous' means 'nearly synchronous'. In these systems each of the individual lower order multiplex streams has an independent clocking system so as a result there will be slight differences in timing of each data stream. To overcome these problems the higher order bit rate of the multiplexed system is slightly higher than the sum of the individual input rates. Any of the extra bits that are not required are filled with what are known as justification bits. Because the timings of all the low order data streams are not exact, it is more difficult to exactly identify the start of a lower-level multiplex bit stream within a high order one. To overcome this problem it is therefore necessary to demultiplex the higher order bit stream in order to access the lower order data stream. This is done using drop-and-insert or add-drop (ADM) multiplexers. An illustration of the use of drop-and-insert multiplexers to access a 2 Mbps data stream at an intermediate location on a 140 Mbps fiber-optic system is shown in Figure 6.8. The multiplex equipment required to insert a small number of channels into a high order multiplex data stream can be complex and expensive. The only way to avoid this is to use synchronous data systems such as SDH or SONET as described in Section 6.15.

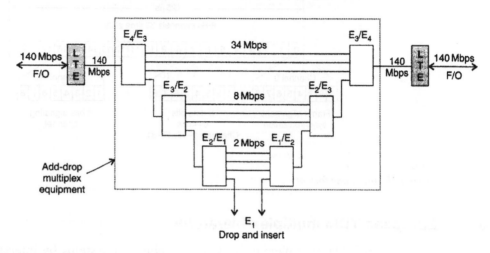

Figure 6.8
Use of drop-and-insert multiplexers

6.5   Switched digital services

The alternative to analog switched services are the digital switched services. The digital switched services are gradually replacing analog services as networks move to all-digital provision. The digital transmission systems as detailed in Section 6.4 are now being connected to digital 4-wire switches to create a switched digital network. In the next sections the following types of digital switched services will be discussed:

- Switched 56
- Integrated services digital network (ISDN-BRI)
- Frame relay (SVC)
- Switched multi-megabit data services (SMDS)
- Asynchronous transfer mode (ATM).

6.6   Switched 56

Switched 56 or global switched digital services is a digitally switched or dial-up service, used primarily in North America, that has the capacity to provide a single data channel of 56 kbps on demand. Different telecommunications companies call the service by different names; some of these are listed in Table 6.6. Switched 56 access lines were some of the first digital circuits installed by the telecommunication carriers. The low cost of switched 56 service relative to digital leased lines makes it cost-effective for supporting sporadic high-speed applications extending into new or existing locations. Although widespread, especially in populous areas, switched 56 circuits are becoming outdated with the implementation of ISDN. The type of connectivity provided by a switched 56 service is ISDN-compatible, with the ability to carry switched digital signals such as video, voice and data. However ISDN services give greater flexibility due to the use of the No. 7 common channel signaling system.

Company	Service Name
Ameritech	PGSDS
AT&T	ACCUNET (GSDS)
Bell Atlantic	Switched 56
Bell South	Accupulse 56
GTE	Switched Data
Nynex	Switchway 56
Pacific Bell	Centrex or CenPath
Southwest Bell	MicroLink 1
US West	Switchnet 56

Table 6.6
Alternative names for switched 56 service

6.6.1 Implementing switched 56 service

To provide dial-up capable digital lines for support of switched 56 requires special CSU/DSUs to be installed on the 56 kbps circuit at the Central Office. The channel service unit (CSU) is a device that is installed between the customer's multiplexer and the service provider's network switch. It manages the physical characteristics of the signal such as the network protection, diagnostics, signal format, clocking and one's density. It is also used here to extract signaling information from the data stream to control the switching of the circuit.

The data service unit (DSU) is installed between the CSU and the user. This unit provides for things such as proper termination specifications, data isolation, and proper mechanical connection specifications. In other words, the data service unit would be the actual physical connection between a user and the data circuit channel service unit. The connections for switched 56 service are illustrated in Figure 6.9.

AT&T use their 4ESSTM switch for 4-wire switching and use the 700 56X XXXX number for switched 56 services. The installation of a Northern Telecomm (Nortel) data unit at the customer's premises is all that is required to support 2-wire switched 56. Once in the PSTN, the individual switched 56 circuits can be routed or switched to any other termination point connected to the PSTN.

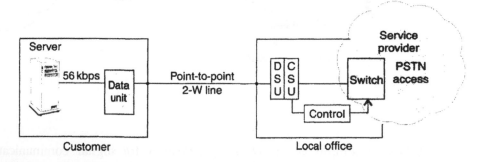

Figure 6.9
Switched 56 service implementation

6.6.2   Applications for switched 56

Switched 56 is useful to use when the amount of data is sporadic and relatively small. It will handle a fairly large amount of data quickly and keep costs down because the connection will be made only when it is actually needed. The switched 56 line is one of the best value dial-up data circuits available today. ISDN lines can handle higher bandwidths than switched 56, but the cost of ISDN has not come down to match the cheaper rates available on switched 56 circuits in most parts of the United States. There is also plenty of interface equipment available to connect switched 56 circuits. ISDN requires more expensive modems and/or interface equipment. When the cost of implementing ISDN comes down, then switched 56 circuits will start to be phased out. Telecommunications solutions for backup and restoration situations or dial-on-demand situations often use switched 56 circuits. With the pricing of switched 56 service based mainly on usage, it is also a viable solution for digital audio and video desktop-conferencing applications.

Switched 56 is a very cost-effective solution because connections are only made when data needs to be transmitted. Further, the costs for terminating equipment, circuit setup and installation fees, and the access charges for switched 56 are much lower than for the newer technologies such as ISDN, frame relay or ATM. There are many vendors who are providing on-demand switched frame relay connectivity over T-1 circuits, but the cost for this type of service exceeds the cost of a switched 56 connection.

6.7   Integrated services digital network (ISDN-BRI)

ISDN was developed to initially utilize the existing copper distribution plant to provide an integrated digital transmission system between the customer and the integrated digital telecommunications network (IDN). ISDN extends 64 kbps digital channels right to the customer and it enables them to carry any combination of digital voice telephony, digital data, telex, facsimile, etc. This is illustrated in Figure 6.10.

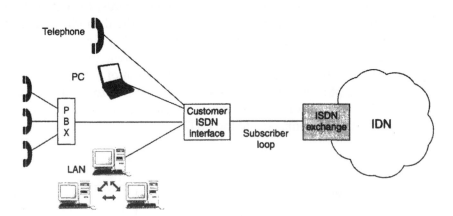

Figure 6.10
Basic ISDN connection

ISDN is simply an interface specification for digital communication between the customer's premises and the first digital switch. Once the connection is made to the digital network, the switches communicate across the network using common channel signaling

(SS No. 7). The network is able to direct the digital traffic to any of the appropriate services anywhere in the network using the SS No. 7 common channel signaling to control the network and provide call management. This is shown in Figure 6.11.

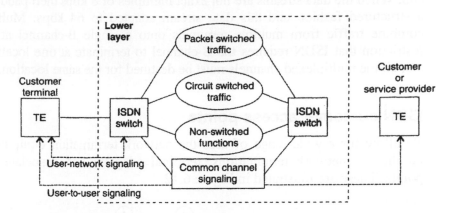

Figure 6.11
ISDN communications architecture

6.7.1   ISDN channel types

ISDN defines two types of user interfaces: the basic rate interface (BRI) primarily for residential customers and the primary rate interface (PRI) for commercial users with greater data requirements. In this section the basic rate interface will be discussed and the primary rate interface is discussed in detail in Section 6.12.

The user access link is made up of combinations of the following user channels:

B-channel   64 kbps. This channel can be used to carry digital data, digitally encoded voice, or mixtures of lower-rate traffic as subchannels to the same destination. These channels can be used on a circuit-switched basis, or they can carry packet-switched data to a packet-switching node using the X.25 protocol, or they can be setup on a semi-permanent basis as a leased line.

D-channel   16 kbps (BRI) or 64 kbps (PRI). This channel is primarily used for carrying the common channel signaling information to control the circuit-switched calls on the associated B-channels at the user interface. The 16 kbps D-channel on the basic rate interface (BRI) can also be used for transmission of low speed data such as telemetry or X.25 packet-switched data.

H-channel   384, 1536 or 1920 kbps. These channels are only used on a primary rate interface (PRI) as discussed in Section 6.12.

The basic rate interface (BRI) consists of two full-duplex 64 kbps B-channels plus a full-duplex 16 kbps D-channel and this configuration is normally referred to as '2B + D'. The total bit rate on a basic access link is 192 kbps due to the additional framing, synchronization and other overhead bits that are required on the channel. This allows simultaneous transmission of voice and several other data services through a single multi-function terminal. Most existing 2-wire local loop plant can support this interface. Depending on your hardware, you can connect up to eight distinct devices to your BRI. This allows you to build a network of devices, phone, data, and video, and use any three of them at the same time.

The B-channel can support devices operating at data rates less than 64 kbps by using rate adaption or multiplexing. Rate adaption works by adapting a single input terminal having a data rate of less than 64 kbps to transmit as a 64 kbps data stream. If the stream is already at exactly 8, 16 or 32 kbps then the first 1, 2 or 4 bits, respectively, of the incoming data are sent in the B-channel, with the remaining bits being set to 1 and discarded at the receiving end. When the data streams are not exact multiples of 8 kbps then padding bits are added in a structured fashion and this data stream adapted to 64 kbps. Multiplexing is used to combine traffic from multiple terminals onto a single B-channel at 64 kbps. Note the restriction that ISDN requires the B-channel to terminate at one location in the PSTN, so that all the multiplexed channels must be destined for the same location.

6.7.2   ISDN customer access points

To allow for a wide range of uses the network termination equipment for ISDN has a number of network terminations, terminal adapters and associated alternative access points. These are illustrated in Figure 6.12.

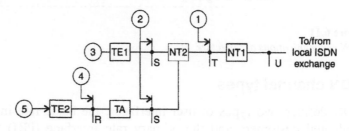

Figure 6.12
ISDN customer access points

The ISDN standards define five groups of functional devices: The NT1, NT2, TE1, TE2 and the TA. After defining the functional devices, the standards define the protocol that each device uses to speak with the devices on either side of it. These protocols are assigned the letters R, S, T and U as identifiers, and are termed reference points. The function of each of these devices is as follows:

Network terminator 1

The network terminator 1 (NT1) is the device that communicates directly with the Central Office switch. The NT1 receives a 'U' interface connection from the phone company, and puts out a 'T' interface connection for the NT2, which is often in the same piece of physical hardware. The NT1 handles the physical layer responsibilities of the connection, including physical and electrical termination, line monitoring and diagnostics, and multiplexing of D- and B-channels. This corresponds to layer 1 in the OSI model.

Network terminator 2

The network terminator 2 (NT2) sits between an NT1 device and any terminal equipment or adapters. An NT2 accepts a 'T' interface from the NT1, and provides an 'S' interface. In most small installations, the NT1 and NT2 functions reside in the same piece of hardware. In larger installations, including all PRI installations, a separate NT2 may be used. ISDN network routers, terminal controllers, LAN adapters and digital PBX's are examples of NT2 devices.

The NT2 handles data-link and network layer responsibilities in ISDN installations with many devices, including routing and contention monitoring.

Terminal equipment 1

Terminal equipment 1 (TE1) devices support the standard ISDN interface and speak the 'S' interface language natively, and can connect directly to the NT devices. Examples of TE1 devices would be an ISDN workstation, an ISDN fax, or a digital telephone. This is shown as access point 3 in Figure 6.12.

Terminal equipment 2

Terminal equipment 2 (TE2) devices are all non-ISDN devices – in fact, every telecommunications device that is not in the TE1 category is a TE2 device. An analog phone, a PC, a terminal with an RS-232 interface, and a FAX are all examples of TE2 devices. To attach a TE2 device to the ISDN network, you need the appropriate terminal adapter. A TE2 device attaches to the terminal adapter through the 'R' interface. This is shown as access point 5 in Figure 6.12.

Terminal adapters

These devices connect a TE2 device to the ISDN network. The terminal adapter (TA) connects to the NT device using the 'S' interface and connects to a TE2 device using the 'R' interface.

Terminal adapters are often combined with an NT1 for use with personal computers. Because of this, they are sometimes referred to as ISDN modems. This is not accurate, because TA devices do not perform analog–digital conversion like modems.

Reference point R

Reference point R (rate) provides a non-ISDN interface between user equipment that is not ISDN incompatible and adaptor equipment. Typically, this interface will comply with an X series or V series ITU-T recommendation. This is shown as access point 4 in Figure 6.12.

Reference point S

Reference point S (system) corresponds to the interface of individual ISDN terminals. It separates the user terminal equipment from the network related communications functions. This is shown as access point 2 in Figure 6.12.

Reference point T

Reference point T (terminal) corresponds to a minimal ISDN network termination at the customer's premises. It separates the network provider's equipment from the user's equipment. This is shown as access point 1 in Figure 6.12.

Reference point U

Reference point U (user) describes the full-duplex data signal on the subscriber's line. This is only used with North American systems where different national ISDN standards are in existence, and this is their network entry point.

An important distinction needs to be made between North American and European practice. In North America both NT1 and NT2 belong to the ISDN customer and the U interface defines the network entry point. In most other administrations, NT1 is considered to be part of the digital network and belongs to the telecom network provider.

6.7.3   User–network interface

The user–network physical interface is illustrated in Figure 6.13. The line to the local ISDN exchange is a single twisted pair supporting duplex transmission for the basic rate access. The hybrid transformer combines the 4 wire transmit and receive paths onto the 2-wire line. The adaptive echo canceler is used to prevent any of the transmitted signal leaking into the received path. The scrambler circuits are used to randomize the bit sequences to avoid correlation between the transmit and receive signals. The S-bus is a 4-wire interface to the terminal equipment or terminal adaptors. It allows up to eight independent devices to communicate using a contention control scheme to timeshare the two B- and the one D-channels.

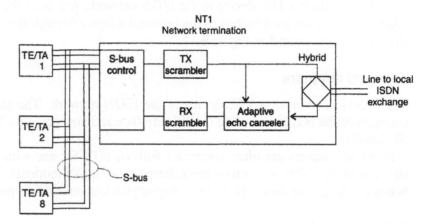

Figure 6.13
User–network interface components

6.7.4   ISDN standards

Standard ISDN is defined by a number of ITU-T I series recommendations as follows:

- I.100 Series – general concepts
- I.200 Series – service capabilities
- I.300 Series – network aspects
- I.400 Series – user–network interfaces.

In the United States, the most widely used ISDN standards are called National ISDN. National ISDN-1 (NI-1) standard covers all basic rate interfaces, National ISDN-2 (NI-2) was to standardize the service features for the primary rate interface, while the NI-3 standard adds new services including; music on hold, enhanced caller ID, interfaces to PCS and frame relay services.

The relationship between the OSI model and the various ISDN standards is illustrated in Figure 6.14.

6.7.5   ISDN circuit switching

ISDN can provide both circuit and packet switching. In the circuit-switching digital networks the 64 kbps B-channel is fully transparent to the network so the user can transmit any kind of data as long as both ends agree on the same protocols. The D-channel is used for controlling the switching of the associated B-channels. At the local exchange the D-channel signaling information is converted into SS No. 7 signaling data using ISDN User part (ISUP), then this information is transferred across the network using the common channel signaling system. In a circuit switched application, the B-channel is switched using only the OSI physical layer (layer 1) functions. This is illustrated in Figure 6.15.

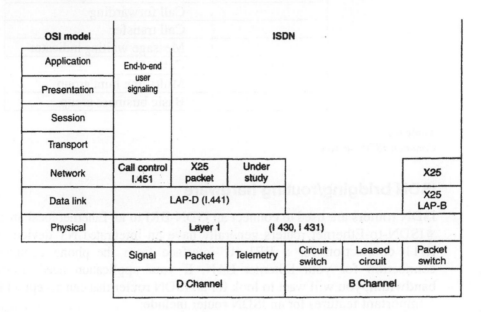

Figure 6.14
ISDN standards and OSI layers

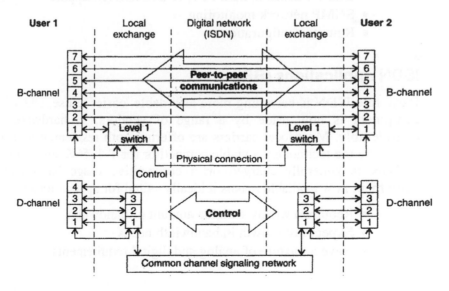

Figure 6.15
Illustration of ISDN circuit-switching connection

6.7.6 ISDN features

ISDN voice features rival the most complicated of PBX systems, but it is unlikely that you will need all of those capabilities. Many new services are available and the exact services you can get in your area depend on the telecom service provider. Table 6.7 lists some of the main ISDN services commonly available.

Category	Service
Voice	Caller ID
	Call hold
	Call conferencing
	Call forwarding
	Call transfer
	Message waiting indicator
Data	Caller ID
	Multi-line hunt group
	Basic business group

Table 6.7
Common ISDN services

6.7.7 ISDN bridging/routing hardware

ISDN routers are used to connect an ISDN-BRI to an Ethernet local area network.

ISDN-to-Ethernet routers generally have an integrated NT device. They provide one RJ11 connection for a BRI 'U'-interface from the phone company, and an RJ45 connection for your Ethernet LAN. If your application needs more than 128 kbps bandwidth, you will want to look for an ISDN router that can accept a PRI interface.

Important features for an ISDN router include:

- Compression support
- Dial-on-demand, MLPPP+, or BONDING support
- SNMP network monitoring
- Remote configuration.

6.7.8 ISDN applications (BRI)

Basic rate ISDN is becoming more affordable, easier to use, and widely available. ISDN equipment is now made by a range of vendors, so hardware prices have reduced considerably. Most local carriers are offering both residential and office ISDN at prices that make the technology a viable competitor for T-1, 56K, and even POTS connectivity.

ISDN is generally charged on a competitive usage basis and the wide range of equipment now available makes it ideally suited for applications that

- Require a widely variable amount of bandwidth
- Expect fast growth in bandwidth needs
- Have a mixture of analog and digital requirements.

6.8 Frame relay

Frame relay is a high-speed packet switch technology for sending information over a wide area network (WAN). A frame relay network is made up of end points, such as PCs, LANs and servers, frame relay access equipment, such as frame relay access devices (FRAD) and routers and network devices such as switches and multiplexers. These are illustrated in Figure 6.16. The frame relay access device delivers frames to the network in the standard format. The network devices route the frame across the network to the appropriate destination user device.

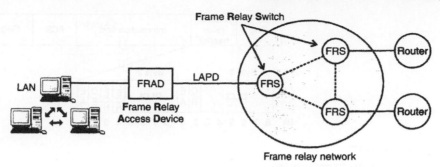

Figure 6.16
Frame relay network

The frame relay network is not a single physical connection between one end user device and another, but the data is transferred over logical paths between the end points, which are called virtual circuits. Bandwidth can be allocated to the path permanently or on a packet by packet basis only when the data is actually transmitted. This makes frame relay networks ideal for the efficient transmission of bursts of data, sharing the bandwidth with a number of similar users.

6.8.1 Virtual circuits

Virtual circuits are two way paths between two end points across the network. There are two kinds of virtual circuit: permanent virtual circuits (PVC) and switched virtual circuits (SVC) as follows:

Permanent virtual circuits

The PVC operate as fixed circuits across the network between the two end points. These are set up as a dedicated point-to-point circuit by the network operator using a network management system. These provide a cost-effective alternative to leased circuits, but they are not available on demand or on a per call basis.

Switched virtual circuits

SVC are able to be established on demand using the SVC signaling protocol. This is a more complex process with call set up needing to track details such as the quantity of data transmitted, addresses, and the amount of bandwidth used for billing purposes. The SVC can be set up dynamically when required, giving the required bandwidth on demand.

6.8.2 Frame structure

The frame relay frame structure is illustrated in Figure 6.17. The various fields are defined as follows:

- Flag: Both opening and closing flags comprise the octet 01111110. A closing flag can serve as the opening flag for the next frame.
- DLCI: Data link connection identifier, a 10-bit virtual circuit number defining the destination on the frame relay network.

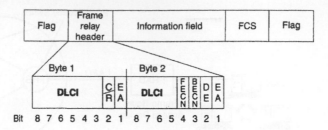

Figure 6.17
Frame relay frame structure

- C/R: Command/response field bit, which is specific to the application.
- EA: Address field extension bit, when additional address octets are required.
- FECN: Forward explicit congestion notification, FECN = 1 is used to advise receiving end system of congestion.
- BECN: Backward explicit congestion notification, BECN = 1 is used to advise all upstream users, including the sender, that congestion is being encountered.
- DE: Discard eligibility, DE = 1 indicates that this frame should be discarded in preference to other frames if congestion is encountered.
- Information: Data from higher level end-to-end protocols, typically 1600 octets maximum for LAN applications to reduce segmentation and reassembly.
- FCS: Frame check sequence, a 16-bit cyclic redundancy check to verify the integrity of the frame contents.

6.8.3   Frame relay operation

Frame relay achieves greater throughput by reducing the amount of processing needed in the network nodes. As a consequence there is no process for error recovery within the frame relay protocols and there is neither acknowledgment of packets nor a guaranteed delivery mechanism. Frame relay packets are often carried on ISDN systems.

Virtual circuit setup

The data link connection identifier (DLCI) has significance on a specific network link and changes as a frame traverses the links along a virtual path through the network. When a virtual circuit is being set up, the local exchange receives the call request packet via the ISDN D-channel. This information is sent to all exchanges along the path by means of the common channel signaling system. Each exchange determines the outgoing link it needs to use to reach the destination address, obtains a free DLCI for that link and makes an appropriate entry in the links routing table between the incoming link DLCI and the outgoing link DLCI. If permanent virtual circuits are used, these routing table entries are made at subscription time.

Frame processing

When a frame is received for data transferred, the frame handler within each exchange reads the DLCI from the frame header, determines the corresponding outgoing link and the new DLCI. This is inserted into the frame header and the frame is queued for transmission on the new outgoing link. If the DLCI is not defined for this frame, the frame is discarded. The use of DLCI is illustrated in Figure 6.18.

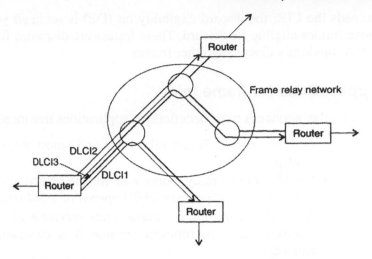

Figure 6.18
Use of data link connection identifiers

Congestion control

The network may be processing multiple calls concurrently over each link and since the frames relating to each call are generated at random intervals, it is possible for an outgoing link to become temporarily overloaded during periods of heavy traffic. The node is able to detect when it is approaching congestion by using internal measurements such as memory buffer usage or the length of the traffic queues. The node is able to give explicit congestion notification by use of the FECN and BECN bits in the frame header. This is illustrated in Figure 6.19. Where node B detects the onset of congestion, it signals by setting the FECN bit in the header of the messages going to node C. At the same time node B sets the BECN bit in the header of all of the messages going back to A.

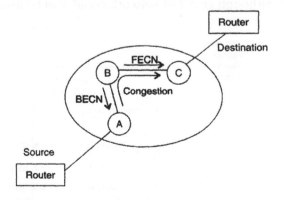

Figure 6.19
Explicit congestion control notification

The frame relay standards require that the user device should reduce its traffic in response to a congestion notification. If the network traffic is not reduced fast enough and an overload occurs then frames will simply be discarded to protect the network. These frames can be discarded at random, but it is preferable to predetermine which frames can be discarded to minimize the disruption to the network. When you subscribe to a frame relay service you need to specify the average information capacity of the required virtual circuit. This is called the committed information rate (CIR). When your frame rate exceeds the CIR, the discard eligibility bit (DE) is set in all your frame headers to make those frames eligible for discard. These frames are discarded first when congestion occurs before randomly discarding other frames.

6.8.4 Applications for frame relay

Frame relay service is most beneficial in applications that meet the following criteria:

- *Multiple sites*:   Frame relay service is most attractive for the connection of multiple sites.
- *Speed*:   Frame relay works well for the provision of service in the data speed range from 56 kbps up to T1/E1 speeds of 1.544 or 2.048 Mbps.
- *Network transparency*:   Frame relay service is ideal for use in a multi-vendor, multi-protocol environment because it is transparent to the actual protocols carried.
- *Traffic variation*:   Frame relay is very effective for handling traffic which is interactive or sent in bursts.
- *Network size*:   Frame relay can be a cost effective choice for linking remote locations because most frame relay tariffs are based only on usage, not circuit distance.

6.9 Switched multi-megabit data services

Switched multi-megabit data services (SMDS) is a connectionless data service designed for networking remote local area networks. In Europe this type of service is known as the connectionless broadband data service (CBDS). The network configuration is illustrated in Figure 6.20, where the two users are located on separate local area networks each of which terminates at the SMDS edge gateway, which is a bridge between the local area network and the digital transmission network. In this case a frame relay network is illustrated although an ATM network could also be used.

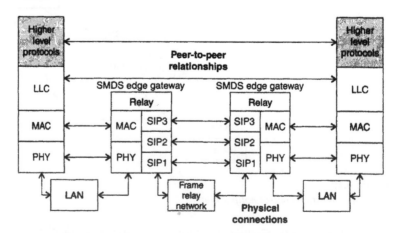

Figure 6.20
Illustration of SMDS protocols

The protocols used with the SMDS are known as the SMDS interface protocols (SIP) levels 1, 2 and 3. The SMDS edge gateway stores the message coming in from the LAN and translates the MAC and PHY layer protocols into the appropriate SIP layers.

The message is then transferred across the frame relay network to the destination edge gateway before being forwarded on to the destination LAN. Other access protocols supported are SMDS data exchange interface (DXI), frame relay and ATM.

6.9.1 SMDS message format

The SMDS service data unit needs to accommodate the data from different types of LAN, and handle the different address formats used in the various types of LAN. The SIP level 3 data packet, or initial MAC protocol data unit (IMPDU) is illustrated in Figure 6.21.

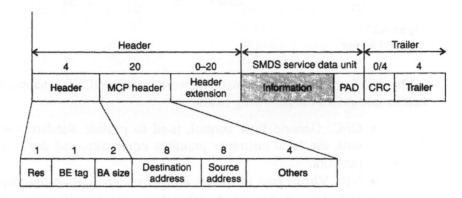

Figure 6.21
Initial MAC protocol data unit

The various fields in this packet are used as follows:

- *Res*: Reserved for future use.
- *BE tag*: Begin–End tag, an 8-bit sequence number used to detect missing frames.
- *BA size*: Buffer allocation, amount of buffer memory required to store the complete IMPDU.
- *MCP header*: MAC convergence protocol header, used to accommodate different LAN formats, and includes the following subfields:

  – *Destination and source addresses*: Up to 64 bits each with appropriate padding, and can include service access point addresses
  – *Others*: The number of PAD octets used, whether a CRC is present and provides for additional subfields to be added in the future.

- *SMDS service data unit*: This has the following subfields:

  – *Information*: Up to 9188 octets to accommodate the largest LAN frame
  – *PAD*: 0–3 octets.

- *CRC*: Cyclical redundancy check to verify the integrity of the frame.
- *Trailer*: The same information as the header.

6.10 Asynchronous transfer mode

Asynchronous transfer mode (ATM) networks use a cell-based transmission system. The data from the input streams is broken down into fixed size transmission units called cells. These cells are multiplexed together for transmission and because the cells have a standard size for all input systems the switching of the different cell streams can be performed at high data rates. ATM uses a cell size of 53-octets, with a payload of 48-octets and a 5-octet header. This is illustrated in Figure 6.22. These small cells provide a more uniform transmission and switching system for constant bit rate traffic as only a short delay is experienced in the segmentation and reassembly of the packets. The packets do not provide error control, so do not need sequence numbers for retransmission.

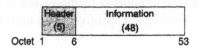

Figure 6.22
ATM cell format

The ATM cell header at the user–network interface (UNI) is shown in Figure 6.23. This header comprises the following fields:

- GFC: Generic flow control, used to provide standardized local flow control with the local customer premises equipment, and does not get sent over the network.
- VPI: Virtual path identifier, 8-bits for identification and routing purposes.
- VCI: Virtual channel identifier, 16-bits for identification and routing purposes.
- PTI: Payload type indicator, indicates the type of information carried in the cell.
- CLP: Cell loss priority, enables the user to specify a preference as to which cells should be discarded. This is the same as the DE (discard eligibility) bit in frame relay.
- HEC: Header error checksum, 8-bit CRC generated by the physical layer on the first 4-octets of the header, to verify its accuracy.

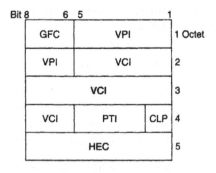

Figure 6.23
ATM cell header

6.10.1   ATM protocol layer functions

A comparison between the OSI model layers and the equivalent layers used with ATM and their functions are shown in Figure 6.24.

6.10.2   ATM virtual paths and channels

An ATM transmission path is made out of virtual paths (VPs) and within each of those virtual paths are virtual channels (VCs). This structure is illustrated in Figure 6.25.

OSI model	ATM		Function
Higher layers	A A L	CS	Convergence
		SAR	Segmentation and reassembly
Data link	ATM		Flow control Cell header Cell VPI/VCI translation Cell multiplex/demultiplex
Physical layer	TC		Transmission convergence
	PM		Bit timing Physical medium

Figure 6.24
ATM potocol layer functions

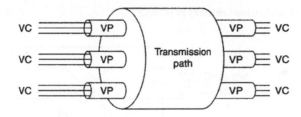

Figure 6.25
ATM transmission paths, virtual paths and virtual channels

Virtual channel

A virtual channel (VC) is used to describe an entity used for unidirectional transport of ATM cells. The virtual channel identifier (VCI) identifies a virtual channel link for a specific virtual path connection (VPC). Each VC link has a VCI assigned when the link is originated and the VC link is terminated when the VCI is removed. Virtual channels are routed by means of a VC switch/crossconnect. This is done by translation of the VCI of the incoming virtual channel links into the VCI values of the outgoing links. This is illustrated in Figure 6.26.

Virtual path

A virtual path (VP) is used to describe a group of virtual channel links having the same end points and share the same VPC. These are identified by a virtual path identifier (VPI). Each VP link has a VPI assigned when the link is originated and the VP link is terminated when the VPI is removed. Virtual paths are routed by means of a VP switch/crossconnect. This is done by translation of the VPI of the incoming virtual channel links into the VPI values of the outgoing links. This is illustrated in Figure 6.27.

6.10.3 ATM services

ATM networks can have the following different service types:

- Constant bit rate (CBR) which operates similar to fixed time division multiplexing but is subject to varying delay, which causes jitter.
- Variable bit rate (VBR) which operates within certain traffic constraints without specific cell flow control from the network.

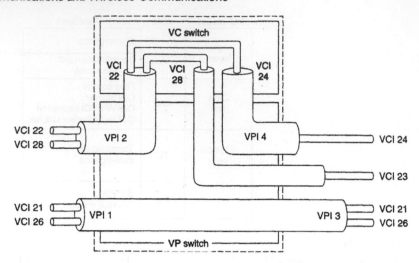

Figure 6.26
Virtual path and virtual channel switching

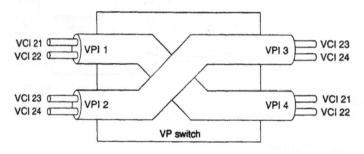

Figure 6.27
Virtual path switching

- Available bit rate (ABR) which is used for data applications. This operates between minimum and peak cell rates, and relies on signals from the network to control cell flow. This results in very low levels of congestion loss.
- Unspecified bit rate (UBR) which provides a best-effort delivery service and can result in congestion loss.

6.11   Digital dedicated circuit (leased) alternatives

Dedicated digital circuits can be obtained from the telecom service provider as various asynchronous (PDH) connections. These have the fixed bandwidth delivered to the customer's premises in various sizes:

- Digital data services (DDS) in the range 2.4–56 kbps
- T-1 (1.544 Mbps) or E-1 (2.048 Mbps)
- Fractional T-1/E-1 which are multiples of 56 kbps
- Greater bandwidth: T-3 at 44.736 Mbps.

The advantage of these services is that the service provider delivers the digital bandwidth to the customer, with all the attendant quality of the digital transmission. The disadvantage is that this bandwidth is fixed, cannot be quickly supplemented if traffic increases, and most services are charged on a flat-rate basis regardless of the circuit usage.

6.11.1 Digital data services (DDS)

DDS provide a high-quality digital point-to-point connection across the telephone network. Most networks incorporate quality of service measurement and a centralized management facility. The advantage of this type of service over the analog data services is that the digital circuit has a better noise performance than analog service. However the service does cost more.

6.11.2 T-1 and T-3 leased services

The North American T-1 services are based on the core DS0 64 kbps channel, as discussed in Section 6.4.1. DS0 signals are aggregated to provide higher levels of bandwidth as follows:

- *T-1*: The T-1 is the most common digital signal size, and provides 1.544 Mbps of aggregated data. A T-1 aggregates 24 DS0 channels, and can handle both voice and data traffic. The T-1 is commonly used to provide high bandwidth data service or trunk lines to a PBX.
- *T-2*: The T-2 is less common in the consumer market. This line uses 96 DS0 channels for a total of 6.312 Mbps. This line was used primarily by the phone companies for high traffic trunk lines, but has been replaced by the T-3 in most applications.
- *T-3*: The T-3 is used for high bandwidth data transmission, with a total throughput of 44.736 Mbps. A DS3 combines 672 DS0 channels.

Customer premise equipment

Customer premise equipment is the physical layer telephony hardware required at each customer's premises to terminate the incoming transmission facilities' circuits. Depending upon the type of circuit, CPE can encompass several different devices.

Channel service units

The channel service unit is the first piece of hardware in the customer premises and it serves the following functions:

- Physically and electrically terminates the connection to the telephone company
- Handles line coding, including bipolar violation correction and B8ZS signaling
- Provides D4 or ESF framing, including error correction and network monitoring.

Data service units

The primary purpose of the data service units (DSU) is to convert the standard unipolar digital signal from the multiplexer into a bipolar signal. The DSU also controls timing and synchronization of the signal.

6.11.3 Fractional T-1 service

This type of service is described in Section 6.4.2. It would be suitable for customers who have large numbers of permanent digital circuits between the sane two locations. A more cost-effective method of circuit provision may be to use some form of statistical multiplexing of the traffic.

6.11.4   E-1 leased services

This is the European equivalent of the T-1 service and is discussed in Section 6.4.3. Fractional E-1 services are also available.

6.12   Integrated services data network (ISDN-PRI)

While the ISDN-BRI was designed as the maximum amount of data that could flow over normal wiring, the primary rate interface (PRI) was designed as the maximum amount of data that could flow over a T-1/E-1 carrier. The main use of the ISDN-PRI today is for large-scale voice services. Many private branch exchange (PBX) units include their own ISDN hardware, and will accept a PRI directly from the telephone company. The ability to easily reallocate trunk lines in the PRI makes it ideal for this function.

The channel structure for the primary rate interfaces is shown in Figure 6.28 for both the North American T1 systems (1.544 Mbps) which can provide 23 B-channels plus one 64 kbps D-channel, and the European E1 systems (2.048 Mbps) which can provide 30 B-channels plus one 64 kbps D-channel using channel 16.

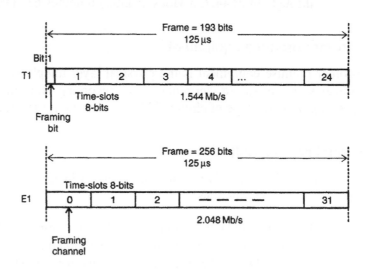

Figure 6.28
PRI frame formats

6.12.1   Aggregating bandwidth with ISDN

One of the most useful features of the ISDN network design is the ability to combine 64 kbps B-channels for applications that require larger amounts of bandwidth. As you add more B-channels to an ISDN installation, you need to choose a method of controlling and combining the channels into one connection. Each B-channel may travel a different path through the network to reach its destination. These different paths may introduce timing differences between data passed over different B-channels. ISDN aggregation techniques can both control multiple B-channel connections, and compensate for timing delays between channels.

6.12.2 The H-channels

When the ITU originally developed the ISDN standards, they included several common B-channel aggregations, called H-channels as follows:

- H0: 384 kbps
- H11: 1536 kbps
- H12: 1920 kbps.

 H-channels can be set up by signaling to the switch, which will manage the opening of the lines and controlling the connection. Because the local switch recognizes the H-channel standard, it is the easiest method of aggregating bandwidth. No additional hardware is needed, and the call is handled just like any other ISDN call. Table 6.8 shows some applications for the common H-channel types.

Bandwidth	Channels	Applications
384 kbps	6	Video conferencing broadcast audio
1.563 Mbps	23	Replaces the T-1
1.920 Mbps	30	Replaces the E-1

Table 6.8
Common H-channel uses

 Another option for applications that need a specific bandwidth allocation that is not addressed by an H-channel is to use special ISDN router hardware that dials groups of B-channels individually.

 The drawback of using an H-channel is that you might waste bandwidth if your application has dynamic needs. The H-channel connection keeps all B-channels open for the entire length of the call, regardless of the demand for bandwidth. To solve this, the industry has developed several methods of dynamically allocating bandwidth.

6.12.3 Inverse multiplexing

Inverse multiplexing is the process of taking one signal and distributing it over several B-channels. The inverse multiplexers are hardware devices that sit on either side of the ISDN connection. These devices are capable of picking up and dropping additional B-channels as bandwidth needs change during the call.

 Because ISDN is priced per B-channel per second, you can save a significant amount by using an inverse multiplexer. On the other hand, the expense of the device, and the additional complexity it causes your installation, may not be worth the line charge savings.

 Each inverse multiplexer has its own method of determining when to pick up or drop a line. Because there is little standardization, it is unlikely that two different inverse multiplexers will be able to talk to each other reliably.

BONDING

Many manufacturers have started adding support for the BONDING protocol to their hardware devices. The BONDING protocol, short for Bandwidth ON Demand

Interoperability Group, was developed in an effort to standardize on a single inverse multiplexing design. BONDING is now defined by the ITU-T Rec. H.221. Many ISDN BRI terminal adapters include BONDING support to allow the end-user to access the full 128 kbps. Because BONDING is a hardware feature, it is fast and efficient, but you need to ensure that both ISDN devices on the connection support it.

MultiLink PPP

The most common method of B-channel aggregation is MultiLink PPP (MLPPP). MLPPP is a similar software solution to BONDING. The only major difference is that standard MLPPP connections cannot dynamically allocate additional B-channels during the call. With standard MLPPP, you choose one or two B-channels at the beginning of the call, and you keep that number of B channels open for the entire duration of the call.

A newer standard (MLPPP+), developed by ASCEND, solves this problem. Calls using MLPPP+ may begin with only one B-channel. When a certain percentage of that B-channel is used, the second B-channel is opened up. When the second B-channel is not needed, it is dropped. MLPPP+ is being widely implemented, and is supported now by several manufacturers.

6.13   Broadband ISDN

Applications which require large amounts of bandwidth for short periods of time are not served efficiently by today's ISDN implementation.

Broadband ISDN (B-ISDN) enables the end-user to get true bandwidth on demand, only paying for the bandwidth used.

B-ISDN supplies bandwidth in excess of 1.544 Mbps, or faster than a T-1. Some applications are shown in Table 6.9.

Bandwidth	Channels	Applications
32.8 Mbps	512	High speed data
44.2 Mbps	690	High speed data
60/70 Mbps	1050	Broadcast video
135 Mbps	2112	Broadcast video, data trunks

Table 6.9
Some B-ISDN applications

B-ISDN will be found in three common forms:

1. *Frame relay*:   Available today in 56 kbps and 1.5 Mbps lines, frame relay service is an example of a B-ISDN solution. Frame relay uses packet switching instead of the circuit switching protocols commonly used in BRIs.
2. *SMDS*:   The SMDS provides packet switched connectivity from 1.5 Mbps to 45 Mbps.
3. *ATM*:   ATM communication uses small, 53-byte 'cells' to transfer data at up to 155 and 622 Mbps. ATM is commonly held as the future direction of B-ISDN services.

The broadband ISDN interface is illustrated in Figure 6.29.

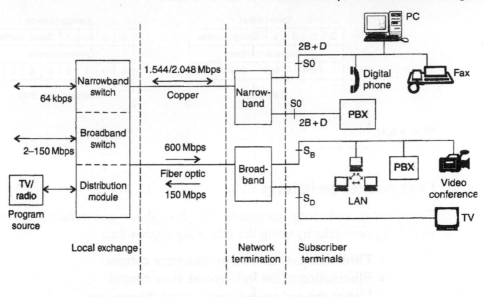

Figure 6.29
Broadband ISDN interface

6.14   X.25 packet switch

X.25 is a set of widely used protocols used to define the interface to the packet switched public data network. The standard provides three levels of protocol:

1. Physical level, X.21 or other standards such as RS-232-C
2. Link level, LAP-B, link access protocol – balanced
3. Packet level, X.25 level 3, provides a virtual circuit service.

The relationship of these protocols in the data frame is shown in Figure 6.30.

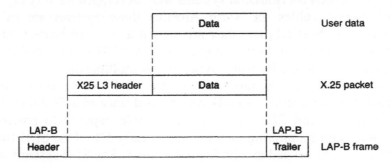

Figure 6.30
X.25 user data and protocol control information

6.14.1   X.25 packet format

Data is transmitted in packets over virtual circuits, which can be set up dynamically as a virtual call or established as a network-assigned permanent virtual circuit. Virtual calls are set up as required and the call cleared on termination. Some X.25 packet formats are illustrated in Figure 6.31.

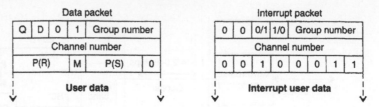

Figure 6.31
Some X.25 packet formats

6.14.2   Fast packet switching

Fast packet switching increases the throughput and reduces the delays on packet switching networks by using the following approaches:

- Elimination of the link-by-link error control
- Elimination of the link-by-link flow control
- Use of the end-to-end error control if necessary
- Use of internal virtual circuits
- Use of hardware switching.

This reduces the amount of overhead in the header of each packet. The frame format is illustrated in Figure 6.32.

Flag	Virtual circuit number	Data	FCS	Flag

Figure 6.32
Fast packet switch frame format

6.15   Synchronous digital hierarchy (SDH/SONET)

Synchronous transmission systems were developed for very high-speed transmission over fiber-optic cables. In North America these systems are called synchronous optical networks (SONET) and they are known as the synchronous digital hierarchy (SDH) in Europe and Japan. In this chapter the term SDH will be used to generically describe both kinds of systems. By using synchronous multiplexing individual tributary signals can be multiplexed directly into a higher rate SDH data stream without need for the complex multiplexing needed on PDH systems as discussed in Section 6.4.6.

SONET (and SDH) have a considerable capacity to provide network management, control and monitoring facilities with nearly 5% of the SDH frame allocated for support of these control and monitoring capabilities. The SDH systems are designed to support all tributary signals used in both the North American (T1) and European (E1) multiplexing hierarchies. SDH systems are able to carry any octet-based data in any protocol, including TCP/IP, frame relay, ATM and SMDS.

6.15.1   SDH signal structure

SDH uses a synchronous signal made up of 8-bit octets organized as a data frame. The basic synchronous transport signal level 1 (STS-1) frame is normally represented as a two-dimensional matrix comprising 90 columns and 9 rows, as depicted in Figure 6.33. The first three columns represent the transport overhead and the envelope capacity is carried in the synchronous payload envelope (SPE) of 87 columns by 9 rows. This STS-1 frame is repeated 8000 times per second and has a bit rate of 51.840 Mbps. These lower-level modules can be multiplexed together to produce STS-N electrical signals. These electrical signals can be converted into optical signals as OC-N where the OC stands for optical carrier. SONET currently supports OC-1, 3, 12, 24, 48 and 192. The SONET and SDH equivalence is shown in Table 6.10.

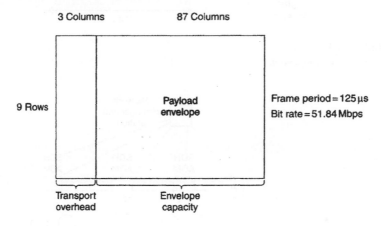

Figure 6.33
STS-1 frame format

SDH	SONET	Bit Rate (Mbps)
	STS-1/OC-1	51.84
STM-1	STS-3/OC-3	155.52
	STS-9/OC-9	466.56
STM-4	STS-12/OC-12	622.08
	STS-18/OC-18	933.12
	STS-24/OC-24	1244.16
	STS-36/OC-36	1866.24
STM-16	STS-48OC-48	2488.32

Table 6.10
Equivalent SDH and SONET systems

6.15.2   SDH system structure

The components in a SDH system are illustrated in Figure 6.34. A section is a single length of transmission cable terminated with section termination equipment, such as a regenerator. A line extends across multiple cable sections and is terminated by line termination equipment, such as multiplexers and switches. A path is an end to end a transmission connection through the complete system and this terminates in path termination equipment at each end. The SPEs are assembled in the SDH multiplexer at one end and demultiplexed in the SDH multiplexer at the other end.

6.15.3   SDH multiplexing equivalence

Table 6.10 shows the equivalent SDH and SONET systems. STM means synchronous transport module.

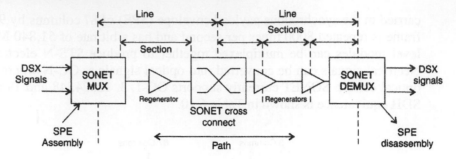

Figure 6.34
SDH system components

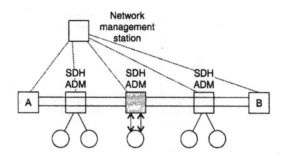

Figure 6.35
SDH multiplexers and network management

6.15.4   SDH system management

All SDH equipments have a software network management agent to communicate with a network management station using communication channels in the overhead bytes of the frame. This information is used to report any malfunction as of the sections, lines or paths. It can also be used to download commands to change the allocation of the payload field in each STM-1 frame to change the bandwidth mix. This management function is illustrated in Figure 6.35.

7

Broadband customer access technologies

Objectives

When you have completed study of this chapter you should be able to:

- List the main technologies available for broadband customer access
- Explain how ADSL systems operate
- Discuss the applications of the various xDSL systems
- Describe the operation of VDSL systems
- Compare the operation of ADSL and G.lite systems
- Describe the application of Ethernet to customer access systems
- Discuss high-speed data access for HFC and FTTC systems
- Discuss the limiting factors in LMDS system design
- Explain suitable applications for Bluetooth systems.

7.1 Overview

The available technologies for the provision of broadband customer access at greater than 1 Mbps include both wired and wireless services. The wired system alternatives include systems operating over the existing POTS copper pairs, known collectively as xDSL systems, the systems used to provide existing cable television services such as hybrid fiber coax (HFC) and their new counterparts such as fiber to the curb (FTTC) and fiber to the home (FTTH). The wireless systems include multi-channel multipoint distribution system (MMDS), local multipoint distribution services (LMDS) and short-range systems such as Bluetooth. This chapter discusses the operation of each of these alternative customer access technologies together with their relative merits and associated performance issues.

7.2 Asymmetric digital subscriber line

Asymmetric digital subscriber line (ADSL) uses existing twisted-pair telephone cable to transmit data at up to 6 Mbps downstream to the subscriber and up to 640 kbps upstream from the subscriber. The basic ADSL system configuration is shown in Figure 7.1.

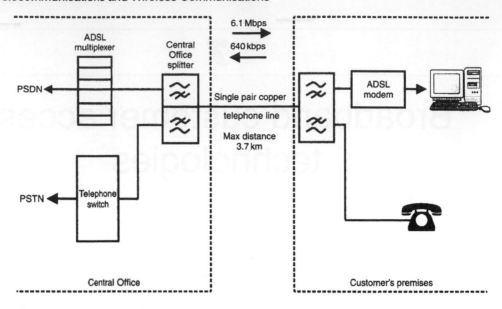

Figure 7.1
Basic ADSL system configuration

ADSL uses standard unconditioned customer access cable pairs and performance primarily depends on the length of the copper circuit, its wire gage, the presence of any bridged taps and the noise environment. Typical ADSL performance for lines without bridged taps is shown in Table 7.1.

Date Rate (Mbps)	Wire Gage (AWG)	Wire Diameter	Distance (km)	Distance (ft)
1.5 or 2.0	24	0.5	5.5	18 000
1.5 or 2.0	26	0.4	4.6	15 000
6.1	24	0.5	3.7	12 000
6.1	26	0.4	2.7	9 000

Table 7.1
Typical ADSL performance

ADSL uses a modem at each end of a twisted-pair telephone line. This creates three information channels: a high-speed downstream channel, a medium speed duplex channel and a POTS voice channel. The POTS voice channel is separated from the digital modem by splitters (filters), to ensure continuous circuit operation in the event of ADSL modem failure. The high-speed downstream channel operates between 1.5 and 6.1 Mbps, while the duplex channel data rate varies between 16 and 832 kbps. The frequency band for ADSL is divided into three portions as shown in Figure 7.2. If echo cancellation is used, it can be seen that the downstream bandwidth can be expanded to overlap the upstream bandwidth. The echo canceler is able to eliminate errors caused by crosstalk from its own signals into adjacent wires within the cable, by subtracting what it transmitted. However errors caused by near-end crosstalk from other circuits cannot be eliminated in this way since the receiver does not know what was transmitted on those adjacent circuits.

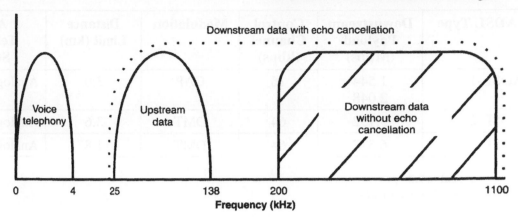

Figure 7.2
ADSL frequency bands

Two different line coding methods are available for ADSL modems: carrierless amplitude phase (CAP) and discrete multitone (DMT). CAP modems use quadrature amplitude modulation (QAM) where the symbols are transmitted as simultaneous combinations of amplitude and phase. In this way up to 1024 discrete line conditions can be created. The carrier signals are suppressed so that only the sidebands are transmitted. Achievement of the highest number of bits carried by each symbol requires low noise transmission channels.

Discrete multitone coding requires the entire bandwidth of the channel to be divided into 256 discrete sub-channels of about 4 kHz. The first six sub-channels are reserved for analog telephony and 24 sub-channels are used for the upstream data. The downstream data uses 222 sub-channels in the absence of echo cancellation, while a further 26 sub-channels can be used with echo cancellation. Each sub-channel uses a distinct sub-carrier which can support the transmission of between 2 and 15 bits per tone using QAM. The receiver automatically adjusts to handle the optimum number of bits depending on the signal-to-noise ratio. When noise is encountered in a particular sub-channel the number of bits per tone is automatically reduced using a procedure known as bit swapping. This is constantly monitored by the exchange of control signals. DMT is considered superior to CAP because it is more flexible, has better noise immunity and automatically optimizes its transmission rate in finer increments: 32 kbps per step for DMT compared to 340 kbps for CAP.

ADSL modems can provide data rates consistent with both the North American and European digital hierarchies and are designed to integrate smoothly with ATM. ADSL modems currently available range from the basic configuration incorporating either 1.5 or 2.0 Mbps downstream data rates with a 16 kbps duplex data channel through to versions providing rates up to 6.1 Mbps with a 64 kbps duplex channel. These are reflected in three types of ADSL service: ADSL-1, ADSL-2 and ADSL-3 as itemized in Table 7.2.

The relevant international standards for ADSL have been established by ANSI T1.413 issue 2 and the ITU-T G.992.1. The basic ADSL standard is ANSI T1.413 issue 1, category 1. Issue 1, category 2 added trellis coding and echo cancellation to enable higher data rates to be transmitted over greater distances. Issue 2 added a rate-adaptive operating mode and provides an optional ATM interworking layer.

A comparison of the performance of the various xDSL systems and EtherLoop and their respective applications are given in Table 7.3.

ADSL Type	Downstream Payload (Mbps)	Control Channel (kbps)	Modulation	Distance Limit (km)	Analog Telephony Support
ADSL-1	1.544 2.048	16	CAP	3.0	Analog
ADSL-2	3.152	64	DMT	3.6	Analog/ISDN/H0
ADSL-3	6.312	64	DMT	1.8	Analog/ISDN/H0

Table 7.2
ADSL service types

System	Max Payload (Mbps)		Distance (km)	Splitter	Application
	Downstream	Upstream			
ADSL	6.144	0.64	3.7	Yes	High speed data
HDSL	1.544–2.048	1.544–2.048	5.0	No	T1 or E1 service access
SDSL	0.192–2.32	0.192–2.32	3.6	No	T1 or E1 service access
VDSL	55	55	0.3	Yes	FTTC/FTTB/FTTN enabling
G.lite	1.5	0.512	5.5	No	Internet access
EtherLoop	5 0.8	5 0.8	1.5 6.5	No	Internet/LAN

Table 7.3
xDSL Comparison

7.3   High-data-rate digital subscriber line

High-data-rate digital subscriber line (HDSL) is essentially a method of providing T1 or E1 service access and is not really a customer access technology. As such it is mentioned here for completeness only. HDSL is defined by ITU-T G.991.1 and ANSI T1E1.4 Tech Report 28. It provides symmetric transmission at speeds of up to 2.048 Mbps of the distances up to 5 km without repeaters or 12 km using repeaters. It operates with two cable pairs per system and utilizes 2 binary, 1 quaternary carrierless amplitude modulation (2B1Q/CAP) line coding, with 80 or 240 kHz modems.

7.4   Symmetric digital subscriber line

Symmetrical digital subscriber line (SDSL) service provides single pair HDSL service over the existing copper circuit between and the Central Office and the customer's premises. Speeds of up to 768 kbps are achievable in both directions for distances up to 12 000 ft or 3.6 km. Some SDSL implementations, such as Lucent's AscendTM, allow two SDSL ports to be combined over two pairs to give a full symmetrical T1 channel capacity of 1.544 Mbps. Figure 7.3 illustrates a full-duplex T1 SDSL configuration.

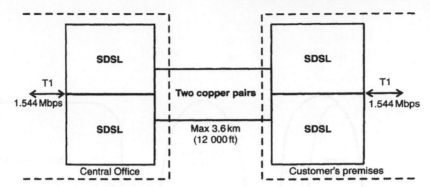

Figure 7.3
Full-duplex T1 SDSL configuration

The relevant international standards for SDSL are being established by ANSI study group T1E1.4 HDSL2 and the ITU-T study group G.shdsl.

7.5 Very-high-speed digital subscriber line

Very-high-speed digital subscriber line (VDSL) transmits high-speed data over a short twisted-pair copper telephone line. The maximum downstream rate being considered is up to 55 Mbps over lines up to 1000 ft or 300 m. Upstream data rates under consideration range from 1.6 Mbps, 19.2 Mbps through to the full downstream rates. Like ADSL, both the data channels will be separated in frequency from the POTS and/or ISDN circuits, to enable service provision on existing cable plant. VDSL is an enabling technology for FTTC installations as discussed in Section 7.9.

VDSL technology is very similar to ADSL, although ADSL has greater problems with the signal dynamic range and attendant crosstalk and is therefore more complex as a result. To enable VDSL to transmit compressed video in real time the transmissions need to incorporate forward error correction (FEC) and the frames need to be interleaved sufficiently to be able to correct all errors caused by impulse noise. Such interleaving introduces delay of approximately 40 times the maximum length correctable impulse. VDSL modems use adaptive equalization techniques, sophisticated synchronization mechanisms and forward error correction to achieve this.

The lower speed implementations of VDSL use adaptive QAM transmission using CAP line coding. Frequency division multiplexing is used to separate the upstream and downstream channels from the POTS and ISDN circuits. Typical channel allocation is shown in Figure 7.4. Echo cancellation will be required for the higher speed later generation systems.

VDSL is a technology capable of providing a full service network over existing telephone plant. Many of the new services being contemplated today can be delivered at speeds below 2 Mbps – such as video conferencing, Internet access, video on demand and remote local area network (LAN) access. VDSL offers more channels than ADSL and can provide for services requiring data rates of more than 2 Mbps – such as digital live television or virtual CD-ROM access. A comparison of ADSL and VDSL data rates vs distance is shown in Figure 7.5.

The relevant international standards for VDSL are being established by the ITU-T G.994.1 in collaboration with ANSI study group T1E1.4, European Telecommunications Standards Institute (ETSI), the Digital Audio-Visual Inter-operability Committee (DAVIC), the ATM Forum and the ADSL Forum.

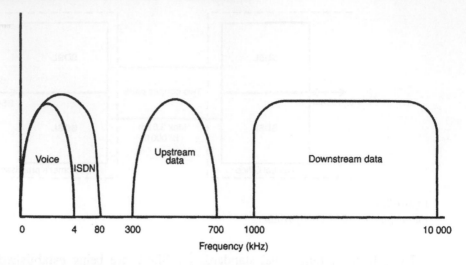

Figure 7.4
Typical VDSL channel allocation

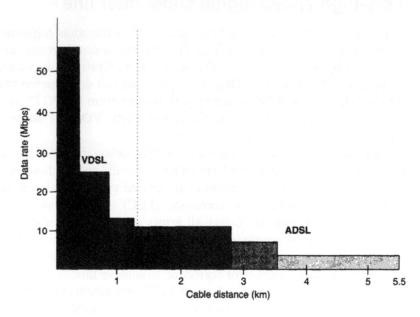

Figure 7.5
Comparison of ADSL and VDSL

7.6   G.lite (ITU-T G.992.2)

G.lite is lower speed, rate adaptive splitterless DSL technology. It uses the same discrete multitone (DMT) line coding as ADSL and provides up to 1.5 Mbps downstream data rate with up to 512 kbps upstream data rate depending on the line conditions and length. The special modulation techniques enable G.lite to provide POTS service on the same line without the use of a splinter at the customer's premises. This facilitates the installation, since rewiring of the customer's telephone is not required. G.lite modems are currently available with a USB interface which provides a plug and play customer installation. The Central Office requires a splitter to separate the voice and data circuits. The data circuits then typically run to a DSL multiplexer with a capacity of up to 8 circuits per card. The typical G.lite system configuration is shown in Figure 7.6.

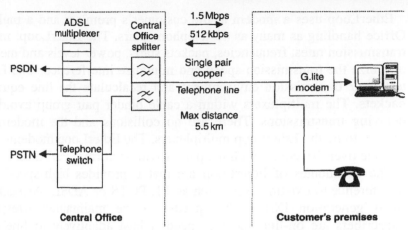

Figure 7.6
Typical G.lite system configuration

G.lite has resolved the noise interference problems of ADSL by transmitting at lower levels and providing greater separation between the upstream, downstream and voice frequencies as shown in Figure 7.7. The G.lite modems communicate with each other at start-up and adapt the speed and power to obtain the best performance under the prevailing line conditions. G.lite modems attempt to start at the highest downstream speed and can typically achieve 1.5 Mbps within 11 s on a good quality line, however on poorer lines the speed is progressively adapted and could take as long as 60 s to work its way down to 160 kbps.

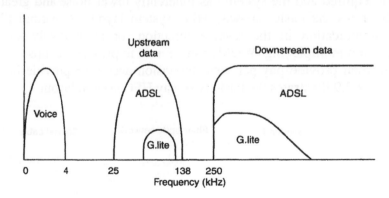

Figure 7.7
Comparison of ADSL and G.lite frequency bands

G.lite provides a cost-effective method of provision of high-speed Internet access for customers up to 18 000 ft or 5.5 km from the Central Office. This is a lower cost alternative to ADSLwhich can provide a continuous data link for Internet traffic while allowing the user to make a simultaneous telephone call on the same line. The G.lite system is defined by the international standard ITU-T G.992.2.

7.7 EtherLoop

EtherLoop is basically Ethernet operating over one existing telephone cable pair. The speed is adaptive from 125 kbps to 6.1 Mbps, depending on the distance. At a distance of 1.5 km from the Central Office data rates of 5 Mbps can be achieved, dropping to 800 kbps at 6.5 km. The data is transmitted in bursts using conventional Ethernet IEEE 802.3 data packets.

EtherLoop uses a modem at the customer's premises and a multiplexer at the Central Office handling as many as 48 telephone pairs. The EtherLoop multiplexers adjust their transmission rates, frequencies, packets starts, power levels and methods of modulation to optimize the transmission speed and minimize interference. The EtherLoop multiplexers test their transmission environment and recalculate the line equalization between data packets. The multiplexers within a cable binder pair group avoid noise interference by delaying transmissions. There are no collisions and the modems only transmit when spoken to by the EtherLoop multiplexers. The EtherLoop modems are splitterless and can operate over the same telephone pair providing normal POTS service.

The advantages of EtherLoop are that it provides high-speed DSL services without interference to existing users such as T1, POTS or ADSL. As such it is considered to be 'next generation DSL'. The speeds can be maintained irrespective of how many subscribers are on-line and the speeds adjust adaptively to line conditions. EtherLoop uses the existing Ethernet packet format and enables Ethernet local area networks to be easily extended up to 21 000 ft or 6.5 km. The total transmission capacity of a 50-pair binder group is approximately 250 Mbps.

7.8   Hybrid fiber coax

Hybrid fiber coax networks are used primarily for the distribution of fixed FDM television programs (CATV) but not necessarily video-on-demand. These networks can be either analog or digital. The analog networks utilize an analog fiber backbone coupled with coax analog broadband distribution amplifiers. The fiber backbone cable simplifies the network structure because of the lower loss in the fiber which means less amplifiers are required and the system has inherently lower noise and greater reliability. Figure 7.8 illustrates the basic one-way HFC system layout. A second fiber is needed to enable communication in the reverse direction, or alternatively some form of wavelength division multiplexing (WDM) can be used to provide the upstream channel. Such systems can then provide pay per view television access or provide high-speed Internet access. Figure 7.9 illustrates the basic two-way HFC system layout.

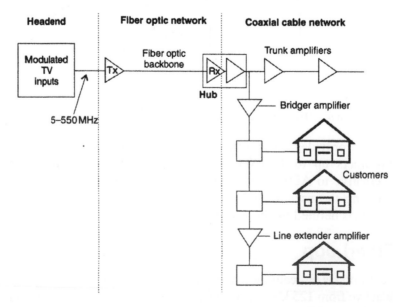

Figure 7.8
Analog HFC system layout (one-way)

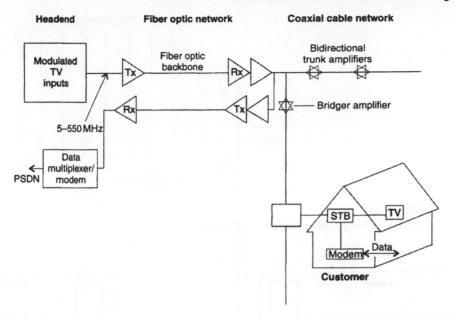

Figure 7.9
Analog HFC system layout (two-way)

Analog HFC can also be used for telephony applications. This uses a home integrated service unit (HISU) which is powered from the cable. The voice channels are handled as primary rate ISDN (2B + D) and 240 64 kbps channels are assigned to a 6 MHz carrier. The frequencies allocated for two way HFC operation are detailed in Figure 7.10. Notice the guard bands used to isolate the upstream data from the downstream TV services.

Frequency

| 30 MHz | 54 MHz | 88 MHz | 144 MHz | 174 MHz | 550 MHz |

Upstream data + voice	Guard band	Low band TV channels 2–6	Downstream data + voice	Mid-band TV channels	Superband TV Channels

Figure 7.10
Analog HFC frequency allocations

Various approaches are used for the transmission of analog CATV signals over the fiber-optic cables. Both systems use intensity modulation of the optical light source. The most common method is to take the 50–550 MHz modulated CATV signal that would have been sent on a coax system, and use that to intensity modulate the fiber light source. The alternative approach is to use frequency modulation of a group of up to eight channels on to a sub carrier and then use this sub carrier to amplitude modulate the intensity of the optical fiber source. In this way a 48-channel distribution system would require six separate fibers. The advantages of this approach are that the system performance is improved (with signal-to-noise ratios in excess of 67 dB achievable), multiplexers are readily available off the shelf, and that smaller numbers of channels can be rolled out incrementally. Figure 7.11 shows the typical allocation of frequencies for an eight-channel per fiber system. Figure 7.12 illustrates a typical FM system block diagram.

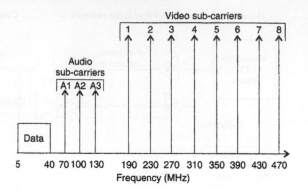

Figure 7.11
Eight-channel per fiber frequency allocations

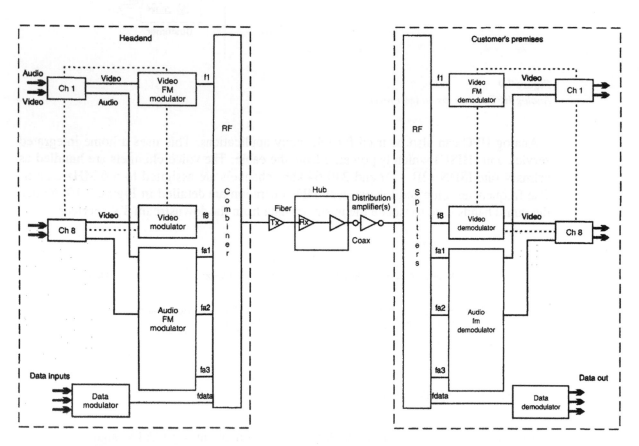

Figure 7.12
Typical FM HFC system block diagram

Digital CATV systems have a major advantage in being able to regenerate signals without accumulation of noise in the digital portion of the network. Consequently the digital trunks can be extended for hundreds of kilometers. Television signals can be transmitted in either compressed or uncompressed format. Uncompressed TV uses sampling rates of around 13.524×106 samples per second which results in data rates of 108.192 Mbps using 8-bit coding or 135.24 Mbps using 10-bit coding. The audio channels (20 kHz) are sampled at 41 880 samples per second using 16-bit PCM resulting in bit rates of 2.68 Mbps for four quadraphonic audio channels. A 16-channel system has a typical bit rate of about 2.38 Gbps. Uncompressed video contains highly redundant information and so its bit error rate requirements are not particularly stringent.

A 4.2 MHz bandwidth NTSC video signal can be compressed to 1.544 Mbps using MPEG-2 compression. By using a 16-QAM modulation scheme three 1.544 Mbps compressed channels can be accommodated within a 6 MHz bandwidth.

The IEEE 802.14 specification addresses the two-way transmission of voice, data, video and interactive data services across international networks. These can include switched data services such as ATM (asynchronous transfer mode), variable length data services such as Ethernet (CSMA/CD), near constant bit rate services such as the MPEG digital video and very low latency data services such as STM (synchronous transfer mode) or virtual circuits. The IEEE 802.14 specification covers the physical (layer 1) and data link (layer 2) layers of the OSI model.

The IEEE 802.14 specification supports two different downstream physical layers in ITU-T recommendation J.83 digital multi-program systems for television, sound and data services for cable distribution: Annex A/C for the European cable systems and Annex B for the North American cable systems. The major difference in the downstream physical layer specifications is in the coding method used for forward error correction (FEC). Type A downstream physical layers use Reed-Solomon block coding, while type B downstream physical layers use a concatenated coding method with outer Reed-Solomon block coding and inner trellis-coded modulation (TCM). Both types of system support two modes of downstream operation utilizing all frequencies above 63 MHz: 64-QAM and 256-QAM, depending on speed.

The standard specifies a tree-structured point-to-multipoint downstream network with a multipoint-to-point bus type of network in the upstream direction. This is shown in Figure 7.13. The multiple 5–42 MHz upstream channels are frequency divisions multiplexed at the fiber node onto a single fiber trunk for transmission back to the headend.

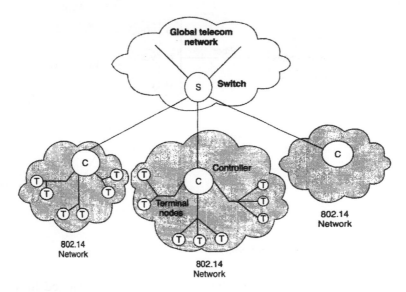

Figure 7.13
HFC CATV network model (IEEE 802.14)

The upstream channels utilize quadrature phase shift keying (QPSK) and 16-QAM modulation techniques. The comparative upstream data rates and modulation rates are detailed in Table 7.4.

Data Rate (Mbps)	OPSK Symbol Rate (Mbaud)	16-QAM Symbol Rate (Mbaud)
0.512	0.256	N/A
1.024	0.512	0.256
2.048	1.024	0.512
4.096	2.048	1.024
8.192	4.096	2.048
16.384	N/A	4.096

Table 7.4
Comparative upstream data and modulation rates

7.9   Fiber to the curb

Fiber to the curb systems provide an intermediate solution between HFC and FTTH. Here the optical fiber cable is terminated in an optical network unit (ONU) which serves up to 30 homes. The connection to the customer is normally provided by coaxial cable but it can also be provided using twisted-pair copper with one of the appropriate DSL technologies. As discussed earlier, VDSL can provide speeds to up to 55 Mbps over distances up to 300 m on conventional twisted-pair copper cables. Alternative names for this technology are fiber to the neighborhood (FTTN) and fiber to the basement (FTTB) provides a similar system designed for serving tall buildings with attendant large vertical cable runs. The system configuration is shown in Figure 7.14.

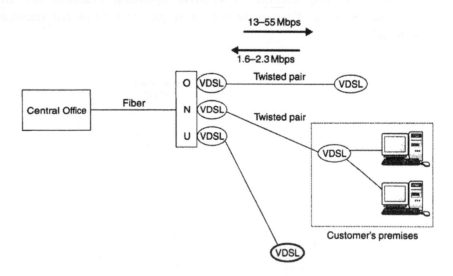

Figure 7.14
Typical fiber to the curb configuration

This provides a cheaper alternative to FTTH, discussed in the Section 7.10, since the cost of the ONU is shared between up to 30 customers. The upstream channel can be provided by using a separate fiber, or alternatively using one of the appropriate DSL technologies over existing copper cables, such as SDSL or VDSL. The provision of appropriate upstream channel speed enables technologies such as switched digital video (SDV) for interactive multimedia services. This is an advantage over HFC systems where the upstream channel has limited capacity for interactive applications. FTTC systems are likely to provide all communications services for the group of residential customers including telephony, interactive video and high-speed Internet access. FTTB systems can provide similar telephony, LANs, WANs and high-speed Internet access for groups of business customers in high-rise buildings.

Data from the optical network unit in the downstream direction will be broadcast to all the terminal equipment in each of the customer's premises or transmitted to a hub which then distributes the data to addressed equipment using cell switching or TDM multiplexing. A typical customer premises network would involve star wiring between each of the items of terminal equipment and a switching or multiplexing hub. An active VDSL network termination is shown in Figure 7.15.

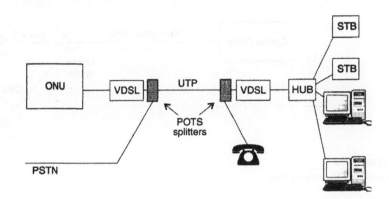

Figure 7.15
Active VDSL network termination

Upstream multiplexing at the ONU is more difficult requiring either some form of control on a common communication channel or use of FDM to establish individual channels per device. On common channels this is done by either token passing between the ONU and terminal equipment, or using a contention system, or both such as contention for unrecognized devices and cell granting for recognized devices. Cell-grant protocols utilize a few bits in the downstream frames to grant access to specific customer premises equipment (CPE) during a specified period. The granted CPE may then send one upstream cell during that period. This involves the transmitter at the CPE turning on, synchronizing with the ONU, sending its cell then turning off. This can be done with a 77-octet frame to transmit a single 53-octet ATM cell. The alternative FDM approach divides the upstream channel into frequency bands for each item of customer premises equipment and thereby avoids the necessity of media access control but has the limitation of restricting the data rate available to any one CPE.

7.10   Fiber to the home

Fiber to the home provides the ultimate customer access technology. Here the fiber-optic cables are extended to the ONU located in the customer's premises. This provides virtually unlimited bandwidth to the customer for all conceivable applications such as provision of video, voice and high-speed data access up to 1 Gbps per customer. FTTH is future-proof, being the only current technology able to meet the demands for such high bandwidth. The basic FTTH architecture is shown in Figure 7.16.

This is a much more expensive option than a HFC because the ONU is required for each customer rather than for a group of up to several hundred customers. FTTH is not cost effective at this time, and is dependent upon advances in technology to provide a more cost-effective bandwidth on fiber-optic cables, and more cost effective ONU technology. The main driver for FTTH will be the requirement for greater bandwidth into the home, since it is the only practical way of providing affordable gigabit data rates to individual customers for the foreseeable future.

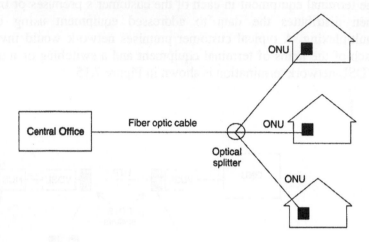

Figure 7.16
Basic FTTH architecture

The simplest method of providing FTTH is to use passive 1:N optical splitters to divide the downstream optical bandwidth roughly equally between the N customers. Similarly passive N:1 combiners are used to consolidate their upstream data onto another fiber. The generic term passive optical network (PON) is used to describe these splitter-based FTTH designs. Alternatively, a single fiber can be used for both directions of transmission using wavelength division multiplexing (WDM). This requires the use of bidirectional splitters and combiners. The splitting of the optical power among many customers in a PON has important optical power budget considerations. Using 16 customers per fiber requires approximately 1/16th of the fiber power to be delivered to each customer resulting in approximately 14 dB lower signal than the equivalent FTTC architecture. The splitting losses can be overcome by the use of optical amplifiers before the passive splitter. The challenge here is to maintain a sufficiently large signal-to-noise ratio at the receiver.

An interesting development involving FTTH is the proposal for the creation of a Gigabit national data grid (GNDG) within the United States. This is intended to overcome the bandwidth and traffic limitations of the Internet and meet all conceivable home and business communications requirements for the foreseeable future. The proposal includes the provision of a Gigabit point of presence (GigaPOP) in each of the 465 congressional districts in the country. By connecting these to local fiber loops and spurs in metropolitan areas, over 90% of the country could be within one mile of a GigaPOP. The proposal also allows the use of the alternative access technologies, already discussed, such as xDSL, coax or wireless to provide service at lower data rates. However interfacing to such solutions may well cost nearly as much as the ultimate fiber to the home infrastructure.

Wavelength Division Multiplexing with appropriate splitters/combiners are the main technologies required to implement such a network. Current commercially available WDM systems are capable of providing bandwidths of 40 Gb per fiber, using 16 wavelengths at 2.5 Gbps each. It is expected that these data rates will reach 100 Terabits within the next 10–15 years.

7.11   Multi-channel multipoint distribution system

Multi-channel multipoint distribution system (MMDS) was developed to distribute video services using microwave systems operating in the 2500–2700 and 2150–2162 MHz frequency ranges. The system operates using the transmitter at the service provider's access node broadcasting line of sight microwave signals to the subscriber's home antenna. This is connected to a set top box (STB) which decodes and decompresses the television signals. This system provides up to 300 high-quality television channels, each of 6 MHz bandwidth, to customers within a 40 mile radius of the transmitter. MMDS enables cost-effective competition for existing cable TV (CATV) providers because it eliminates the necessity of installing a fiber-optic or coax cable network throughout the service area.

MMDS can also be used to provide data services for Internet access.

7.12   Local multipoint distribution system

Local multipoint distribution system (LMDS) uses millimeter wave radio transmission in the Ka-band (28 GHz) to transmit voice, data and video signals within cells of up to 3 miles or 5 km in diameter. LMDS can deliver services ranging from one-way video distribution and telephony to fully interactive switched broadband multimedia applications such as video-on-demand, interactive video, video conferencing and high speed Internet access. The basic LMDS configuration is shown in Figure 7.17.

LMDS can also handle thousands of telephone channels. For example the Texas Instruments 'MulTIpoint' system can provide up to 224 digital video channels and 16 000 telephone channels per residential node. For business applications the same system can provide 192 T1 circuits and up to 4608 DS0 circuits per business node.

LMDS is defined by the IEEE 802.16 standard. Among the various LMDS systems available in the marketplace are 'MulTIpoint' from Texas Instruments and also WirelessMAN™.

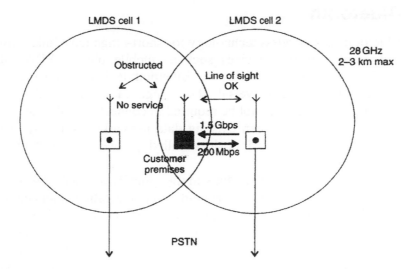

Figure 7.17
Basic LMDS configuration

The attractions of LMDS are the large bandwidth, the low installation costs and ease of deployment. The major problems in using such high frequencies are the fast signal attenuation and the frequency selective attenuation often caused by absorption of the signal in the water molecules in heavy rainfall. These problems can be overcome by ensuring adequate signal strength at the receiver by careful cell design. The cell size is dependent on having a direct line of sight radio path between the transmitter and receiver. Where a direct line of sight path is not available from one transmitter, it is necessary to receive service from another overlapping transmitter. By using appropriately engineered digital signals with large fade margins, the systems have a robust noise tolerance and are less susceptible to the effects of signal absorption due to local rainfall and wet foliage. The heights of the transmitting and receiving antennae are also significant with better reception achieved by higher antenna positioning. Typical antenna size is about 6 in. (150 mm) square.

Tests have shown that excessive path loss is the most serious propagation impairment for operational LMDS systems. At present practical systems appear to be limited to about 2 km path lengths. The other factors affecting signal reception quality include attenuation by vegetation and obstruction of the radio paths by adjacent buildings. Where multi-party propagation exists, delay spread becomes a major impediment.

LMDS provides asymmetrical two-way transmission operating at up to 1.5 Gbps downstream and 200 Mbps upstream. Under heavy system loading downstream data rates of 7 Mbps with 1 Mbps upstream per household are sustainable. This is sufficient to view two movies simultaneously. CellularVision are currently delivering the equivalent of 49 channels of CATV to parts of New York City using LMDS. They plan to introduce two-way services such as video-on-demand and high-speed Internet access starting at 550 kbps. The uplink channels can be inserted between the video channels using reverse polarization.

LMDS has considerable potential for use in countries without a modern telecommunications infrastructure to provide wireless broadband communication. Many countries are considering the allocation of radio spectrum for such broadband applications. A 23-GHz broadband network was installed in Kobe, Japan, to quickly restore CATV service after their devastating earthquake.

7.13   Bluetooth

Bluetooth is a wireless technology for short-range radio links between various computing and communication devices such as mobile phones, PDAs, desktop computers, fax machines, printers, joysticks, etc. Bluetooth technology enables all of these devices to be connected by one universal short-range radio link. This means that many of the proprietary cables will not be required to connect these devices. Bluetooth also provides a universal bridge between existing data networks and a way of providing small ad hoc groupings of connected devices called piconets, without requiring the use of fixed network infrastructure.

Bluetooth operates at 1 Mbps in the globally available 2.4 GHz ISM radio band which ensures that it will have worldwide communication compatibility without the need for licensing. Frequency hopping transceivers are used to ensure protection from interference, fading and for security of data. Forward error correction (FEC) is used to reduce the impact of noise on the transmissions. The Bluetooth radio is built into a small micro chip. The radio modules avoid interference from other signals by hopping to a new frequency after transmitting or receiving a data packet. Bluetooth is more robust than other radio systems operating in the same band because it uses shorter packets and hops faster between frequencies. The maximum frequency hopping rate is 1600 hops per second. The nominal link range is 10 cm–10 m but can be extended to more than 100 m by increasing the transmitter power.

The Bluetooth baseband protocol utilizes both circuit- and packet-switching. Slots can also be reserved for synchronous packet transmission. A packet will normally be sent in a single slot, but may be extended to cover up to five slots. Bluetooth can also support an asynchronous forward channel of up to 721 kbps in either direction with a return channel of 57.6 kbps. The system can alternatively support a 452.6 kbps transmission in both directions. The system will also support up to three synchronous voice channels of 64 kbps or one channel providing simultaneous synchronous voice transmission plus asynchronous data. The voice channels utilize the continuous variable slope delta (CSVD) modulation technique, which has proven very robust for handling damaged or dropped voice samples. Voice packets with errors or missing packets are never retransmitted. At bit error rates as high as 4% the speech is quite audible but has increased background noise.

The Bluetooth air interface is based on a nominal antenna power of 0 dBm to comply with the FCC rules for the ISM band. Spread spectrum operation has been added to enable greater distances up to 100 m to be achieved at power levels up to 100 mW. This makes it attractive for use within the home, for example. This uses up to 79 hops at one MHz intervals starting from 2402 GHz through to 2480 GHz.

The Bluetooth network topologies includes both point-to-point and point-to-multipoint connections. With the current Bluetooth specification, up to seven slave devices can be set to communicate with a single master, this becomes a piconet. The members of one piconet are determined by means of software controls and identity codes in each microchip, so that only the correct members can communicate. All members of the same piconet have priority synchronization and utilize the same frequency hopping sequence. When several piconets are linked together each uses a different frequency hopping sequence, but other devices can be set to enter the communication at any time. The adhoc interconnection of several piconets is known as a scatternet. This is illustrated in Figure 7.18.

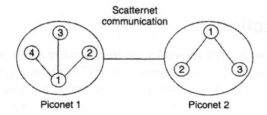

Figure 7.18
Bluetooth piconets and scatternet

8

Local and wide area networks

Objectives

After completion of this chapter, you will be able to:

- Describe the various LAN topologies as well as their advantages and disadvantages
- Describe various media access methods, specifically CSMA/CD, CSMA/CA, token passing and polling
- Describe various LAN standards with emphasis on IEEE 802.3, IEEE 802.5, IEEE 802.11 and ANSI X3T9.5
- Describe the functionality of the following network devices: repeaters, bridges, routers, switches and gateways
- Describe the basic characteristics of LANs, WANs, MANs, VLANs and VPNs.

8.1   Introduction

Local and wide area networks are traditionally associated with data communications only. However, with the convergence of PSTN technologies and LAN/Internet technologies, LANs and WANs are used more and more to carry voice traffic over and above data traffic. The topics of convergence and voice over IP (VoIP) will be dealt with in more detail in Chapter 9.

Because of this phenomenon, a basic understanding of LAN/WAN technology is indispensable. This chapter will deal with generic LAN attributes such as topologies and access methods, and will discuss a few LANs with an emphasis on Ethernet as the dominant technology.

8.2   LAN topologies

8.2.1   Broadcast vs point-to-point topologies

Networks consist of several nodes (workstations, printers, PLCs, etc.) interconnected via some sort of medium. The way the nodes are interconnected is known as the topology. There are many possible topologies but they all form in one of two categories, broadcast and point-to-point.

Broadcast topologies are those where the messages ripple out from the transmitter to reach all nodes. There is no active regeneration of the signal by the nodes and thus signal propagation is independent of the operation of the network electronics. This then limits the size of such networks. Figure 8.1 shows an example of a broadcast topology.

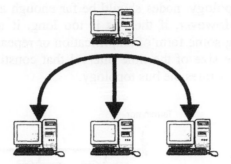

Figure 8.1
Broadcast topology

In a point-to-point communications network, however, each node is communicating directly with only one other node. That node may actively regenerate the signal and pass it on to its nearest neighbor. Such networks have the capability of being much larger. Figure 8.2 shows some examples of point-to-point topologies.

Figure 8.2
Point-to-point topologies

8.2.2 Logical vs physical topologies

A logical topology defines how the elements in the network communicate with each other and how information flows through the network. A physical topology, on the other hand, defines the wiring layout for a network. This specifies how the elements in the network are interconnected electrically. This arrangement will determine what happens if a node on the network fails. Physical topologies fall into three main categories namely bus, star and ring. Combinations of these are used to form hybrid topologies to overcome weaknesses or restrictions in one or other of these three basic topologies.

It will be seen that in many cases, the physical and logical topologies are the same. There are, however, instances e.g. IBM token ring where the physical topology resembles a star, but the logical topology resembles a ring.

8.2.3 Bus (multidrop) topology

Physically, a bus describes a network in which all nodes are connected to a common single communication channel or 'bus'. This bus is sometimes called a backbone, as it provides the spine for the network. Every node can hear each message packet as it goes past.

Logically, a bus is distinguished by the fact that packets are broadcast and every node gets the message at the same time. Transmitted packets travel in both directions along the bus, and need not go through the individual nodes, as in a point-to-point system. Rather, each node checks the destination address that is included in the packet to determine whether that packet is intended for the specific node. When the signal reaches the end of the bus, an electrical terminator absorbs the energy to keep it from reflecting back again along the cable and interfering with other messages already on the bus. Both ends of a bus cable must be terminated in this way, so signals are removed from the bus when they reach the end.

In a bus topology, nodes should be far enough apart so that they do not interfere with each other. However, if the bus is too long, it may be necessary to boost the signal strength using some form of amplification or repeater. The maximum length of the bus is limited by the size of the time interval that constitutes 'simultaneous' packet reception. Figure 8.3 illustrates the bus topology.

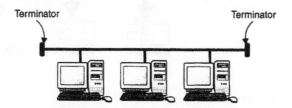

Figure 8.3
Bus topology

Advantages

A bus topology offers the following advantages:

- A bus uses relatively little cable compared to other topologies, and arguably has the simplest wiring arrangement.
- Since nodes are connected by high impedance tappings across a backbone cable, it is easy to add or remove nodes from a bus. This makes it easy to extend a bus topology.
- Architectures based on this topology are simple and flexible.
- The broadcasting of messages is advantageous for one-to-many data transmissions.

Disadvantages

These include the following:

- There can be a security problem, since every node may see every message, even those that are not destined for it.
- Diagnosis/troubleshooting (fault isolation) can be difficult, since the fault can be anywhere along the bus.
- There is no automatic acknowledgment of messages, since messages get absorbed at the end of the bus and do not return to the sender.
- The bus cable can be a bottleneck when network traffic gets heavy. This is because nodes can spend much of their time trying to access the network.

8.2.4   Star (hub) topology

In a physical star topology multiple nodes are connected to a central component, generally known as a hub. The hub usually is just a wiring center, that is, a common termination point for the nodes. In some cases, the hub may actually be a file server (a central computer that contains a centralized file and control system), with all the nodes attached directly to the server. As a wiring center, a hub may in turn be connected to a file server or to another hub.

All packets going to and from each node must pass through the hub to which the node is connected. The telephone system is the best known example of a star topology, with lines to individual customers coming from a central telephone exchange location. An example of a star topology is shown in Figure 8.4.

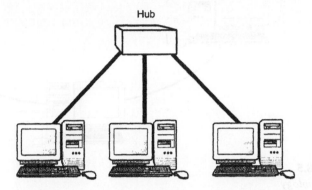

Figure 8.4
Star topology

Advantages

- Troubleshooting and fault isolation is easy.
- It is easy to add or remove nodes and to modify the cable layout.
- Failure of a single node does not isolate any other node.
- The inclusion of a central hub allows easier monitoring of traffic for management purposes.

Disadvantages

- If the hub fails, the entire network fails.
- A star topology requires a lot of cabling.

8.2.5 Ring topology

As a logical topology, a ring is distinguished by the fact that packets are transmitted sequentially from node to node, in a pre-defined order, in a point-to-point system. As a physical topology, a ring describes a network in which each node is connected to exactly two other nodes. Nodes are arranged in a closed loop, so that the initiating node is the last one to receive a packet (Figure 8.5).

Information traverses a one-way path, so that a node receives packets from exactly one node and transmits them to exactly one other node. A packet travels around the ring until it returns to the node that originally sent it. Each node acts as a repeater, boosting the signal before sending it on. Each node checks whether the packet's destination address matches its own. When the packet reaches its destination, the destination node accepts the message, then sends it back to the sender, to acknowledge receipt.

Since ring topologies use token passing to control access to the network, the token is returned to sender with the acknowledgment. The sender then releases the token to the next node on the network. If this node has nothing to say, the node passes the token on to the next node, and so on. When the token reaches a node with a packet to send, that node sends its packet. Physical ring networks are rare, because this topology has considerable disadvantages compared to a more practical star-wired ring hybrid, which is described later.

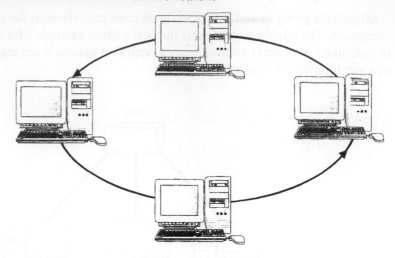

Figure 8.5
Ring topology

Advantages

- A physical ring topology has minimal cable requirements.
- No wiring center or closet is needed.
- The message can be automatically acknowledged.
- Each node can regenerate the signal.

Disadvantages

- If any node goes down, the entire ring goes down.
- Diagnosis/troubleshooting (fault isolation) is difficult because communication is only one-way.
- Adding or removing nodes disrupts the network.
- There will be a limit on the distance between nodes.

As well as these three main topologies, some of the more important variations will now be considered. These are just variations, and should not be considered as topologies in their own right.

8.2.6   Star-wired ring topology

A star-wired ring topology is a hybrid topology that combines features of the star and ring topologies. Individual nodes are connected to a central hub, as in a star network. Within the hub, however, the connections are arranged into an internal ring. Thus, the hub constitutes the ring, which must remain intact for the network to function. The hubs, known as multistation access units (MAUs) in IBM token ring network terminology, may be connected to other hubs. In this arrangement, each internal ring is opened and connected to the attached hubs, to create a larger, multihub ring.

The advantage of using star wiring instead of simple ring wiring is that it is easy to disconnect a faulty node from the internal ring. The IBM data connectors on the hub is specially designed to close the circuit if an attached node is disconnected physically or electrically. By closing the circuit, the ring remains intact, but with one less node. The IBM token ring network is the best-known example of a star-wired ring topology. In token ring networks with dual redundant rings, a secondary ring path can be established and used if part of the primary path goes down. The star-wired ring is illustrated in Figure 8.6.

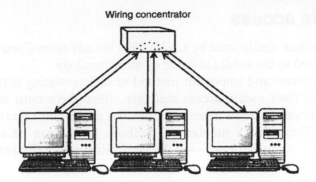

Figure 8.6
Star-wired ring

Advantages

- Troubleshooting, or fault isolation, is relatively easy.
- The modular design makes it easy to expand the network, and makes layouts extremely flexible.
- Individual hubs can be connected to form larger rings.
- Wiring to the hub is flexible.

Disadvantages

- Configuration and cabling may be complicated because of the extreme flexibility of the arrangement.

8.2.7 Distributed star topology

A distributed star topology is a physical topology that consists of two or more hubs, each of which is the center of a star arrangement (Figure 8.7). A good example of such a topology is an ARCnet network with at least one active hub and one or more active or passive hubs. The 100VG AnyLAN utilizes a similar topology.

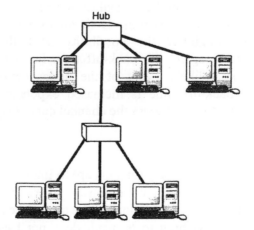

Figure 8.7
Distributed star topology

8.3   Media access

The various media used by LANs have already been discussed in Chapter 2. All nodes are connected to the media in some or other topology.

A common and important method of differentiating between different LAN types is to consider their media access methods. Since there must be some method of determining which node can send a message, this is a critical area that determines the efficiency of the LAN. There are a number of methods which can be considered, of which the most common in current LANs are the contention, token passing and polling methods.

8.3.1   Contention systems

This is the basis for a first-come-first-served media accesses method. It operates in a similar manner to polite human communication. We listen before we speak (at least most people do), deferring to anyone who already is speaking. If two persons start speaking at the same time, they recognize that fact and both stop, before restarting their messages again a little later. In a contention-based access method, the first node to seek access when the network is idle will be able to transmit.

There are two variations on the theme, namely CSMA/CD and CSMA.

CSMA/CD

Contention is at the heart of the carrier sense multiple access/collision detection access method used in the IEEE 802.3 and Ethernet V2 networks.

Carrier sense means that a node wishing to transmit a message listens out on the medium to ensure there is no 'carrier' present. The signaling method used on baseband type systems such as Ethernet do not use a carrier as in broadband systems, but simply encode the data e.g. in Manchester format. There is thus no carrier involved, so the node wishing to transmit simply checks for the presence or absence of a signal on the medium. Once the sender confirms the absence of a signal, it starts transmitting.

The length of the channel and the finite propagation delay of the signal means that there is still a distinct probability that more than one transmitter will attempt to transmit at the same time, as they both will have heard no 'carrier'. It is also theoretically possible for two nodes in close proximity to listen out, and start transmitting at exactly the same time. The collision detection logic on both nodes will detect such a collision (simply by detecting that there is another signal on the medium over and above the one being sent) and terminate the transmissions. Both nodes will back off and attempt to repeat the procedure. To minimize the possibility of both nodes attempting a simultaneous transmission once again, the back-off time involves a calculation involving a random number. In the unlikely event of the nodes actually being involved in yet another collision, the back-off time is increased exponentially. The system is a probabilistic system, since the time to access the channel cannot be ascertained in advance.

Advantages

- The system is easy (and cheap) to implement in hardware/firmware.
- Each node acts autonomously, hence nobody is 'in charge' and there is no common mode of failure of the access method.
- A node wishing to transmit does not have to 'wait its turn'. It can grab the medium as soon as possible.
- It works well in the case of many hosts wishing to transmit short messages.

Disadvantages

- In its basic form, there can be no prioritizing mechanism. A node with an urgent message has the same chance of accessing the medium as one that has not. In Ethernet systems for industrial applications this problem has been addressed by modifying the basic Ethernet frame and including a 3-bit priority field, which allows for eight different priority levels. Unfortunately, this calls for the use of special switching hubs to interpret this field, and to give preference to certain nodes accordingly.
- Basic Ethernet (without the priority feature) is said to be non-deterministic. This means that although access could be quick, it is not possible to guarantee (i.e. mathematically calculate) the worst-case access time for a given node – something that concerns control engineers.
- Because of the increase in collisions and the as traffic increases, networks using this access methods should not be heavily loaded. The recommended maximum loading for Ethernet networks in control applications is 3% whilst for office applications it is 30%.
- It should nevertheless be pointed out that despite Ethernet's use of CSMA/CD, Ethernet networks using full-duplex switches, do not experience collisions at all.

CSMA/CA

Carrier sense multiple access/collision avoidance is a contention scheme used on IEEE 802.11 and other wireless protocols. The sender first broadcasts an 'rts' (request to send) frame. This frame contains information regarding the length of the intended transmission, so that others know how long to wait (or how long not to wait). If a receiver is willing to accept the message, it broadcasts a 'cts' (clear to send) frame. The 'cts' frame also contains the information regarding the intended message. The sender, upon receiving the 'cts', transmits the message.

Any other machine receiving the 'cts' frame knows that it is close to the receiver, and should not transmit within the time interval specified within the 'cts' frame. Any other machine receiving 'rts' but not 'cts' is sufficiently far from the receiver that it can send without interfering.

8.3.2   Token passing

Token passing is a deterministic media-access method in which a token is passed from node to node, according to a predefined sequence. A token is a special packet, or frame, consisting of a signal sequence that cannot be mistaken for a message. At any given time, the token can be either available or in use. When an available token reaches a node, that node can access the network for a maximum pre-determined time, before passing the token on.

This deterministic access method guarantees that every node will get access to the network within a given length of time. This is in contrast to a probabilistic access method (such as CSMA/CD), in which nodes check for network activity when they want to access the network, and the first node to claim the idle network gets access to it. Network architectures that support the token passing access method include token bus, ARCnet, FDDI, and token ring.

To transmit, the node first marks the token as 'in use', and then transmits a data packet, with the token attached. In a ring topology network, the packet is passed from node to node, until the packet reaches its destination. The recipient acknowledges the packet by sending the message onwards until it reaches the sender, who then sends the token on to the next node in the network (Figure 8.8).

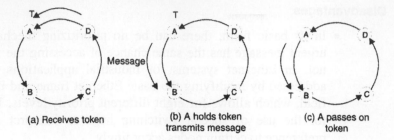

Figure 8.8
Token passing

In a bus topology network, the next recipient of a token is not necessarily the node that is physically nearest to the current token passing node. Rather, the next node is a logical neighbor as determined by some pre-defined rule. The actual message is broadcast on to the bus for all nodes to 'hear'. For example, in an ARCnet or token bus network, the token is passed from a node to the node with the next lower network address.

Networks that use token passing generally have some provision for setting the priority with which a node gets the token. Higher level protocols can specify that a message is important and should receive higher priority.

A token ring network requires an active monitor (AM) and one or more standby monitors (SMs). The AM keeps track of the token to make sure it has not been corrupted, lost, or sent to a node that has been disconnected from the network. If any of these things happen, the AM generates a new token, and the network is back in business. The SM makes sure the AM is doing its job and does not break down and get disconnected from the network. If the AM is lost, one of the SMs becomes the new AM, and the network is again in business. These monitoring capabilities make for complex circuitry on network interface cards that use this media access method.

Advantages

- Because each node gets its turn within a fixed period, deterministic access methods are more efficient on networks that have heavy traffic. With such networks, nodes using probabilistic access methods spend much of their time competing to gain access and relatively little time actually transmitting data over the network.
- Since only the node with the token can transmit, there can never be any collisions.

Disadvantages

- The fact that media access is deterministic does not necessarily make it fast.
- The software token can get lost, duplicated or corrupted. The software required to generate and maintain the token is substantial.
- If the token gets lost or corrupted, no node can communicate.

8.3.3 Polling

Polling refers to a process of checking stations on the network in some defined order, to see whether the polled element wants to transmit. In roll-call polling, the polling sequence controlled by the central element. In contrast, in hub polling, each element simply polls the next element in the sequence.

In LANs, polling provides a deterministic media-access method in which the server or central hub polls each node in succession. In some systems, the polling is done by means of software messages, which could slow down the process. In order to overcome this problem, systems such as 100VG AnyLAN employ a hardware polling mechanism which uses voltage levels to determine whether a node wants to be serviced. HP calls this demand priority processing.

Advantages

- This is also a deterministic method, which is preferable in situations where a guaranteed access time to the medium is important.
- Demand priority processing distinguishes between time-critical frames such as video, and non-time-critical frames carrying data, and handles the former with a higher priority.

Disadvantages

- The process is controlled by a one or more central devices. If any one fails, all nodes connected to that device cannot transmit.

8.4   LAN standards

The various LAN standards essentially occupies layers 1 and 2 (physical and data link layers) of the OSI model. The LAN interface to a host (computer) is normally implemented as a network interface card (NIC). Depending on the type of computer, the card may be an ISA, PCI, PCMCIA or similar card. The data link layer protocols are implemented in firmware, i.e. on a ROM on the card.

The Institute of Electrical and Electronic Engineers in the United States has been given the task of developing standards for local area networking under the auspices of the IEEE 802 committees. Once a draft standard has been agreed and completed, it is passed to the International Standards Organization (ISO) for ratification. The corresponding ISO standard, which is generally internationally accepted, is given the same committee number as the IEEE committee, with the addition of an extra '8' in front of the number i.e. the IEEE 802 committees are equivalent to the ISO 8802 committees.

Not all LAN standards are IEEE standards, though. The fiber distributed data interface (FDDI), for example, has been standardized as ANSI X3T9.5 or ISO 9314.

The IEEE committees, consisting of various technical study and working groups, provide recommendations for various features within the networking field. Each committee is given a specific area of interest, and a separate subnumber to distinguish it. The main committees and the standards that they are working on are described below.

IEEE 802.1 High level interface (HILI)

The HILI subcommittee is concerned with issues such as high level interfaces, internetworking and addressing. There are a series of subcommittees, such as:

- 802.1B: LAN management
- 802.1D: Local bridging
- 802.1E: System load protocol
- 802.1F: Guidelines for layer management standards
- 802.1G: Remote MAC bridges
- 802.1I: MAC bridges (FDDI supplement).

IEEE 802.2 Logical link control

This is the interface between the network layer and the specific network environments at the physical layer. The IEEE has divided the data link layer in the OSI model into two sublayers – the media access MAC sublayer, and the logical link control layer (LLC). The logical link control protocol is common for all IEEE 802 standard network types. This provides a common interface to the network layer of the protocol stack.

IEEE 802.3 CSMA/CD

The carrier sense, multiple access with collision detection type LAN is commonly – but strictly speaking incorrectly – known as an Ethernet. Ethernet refers to the original DEC/INTEL/XEROX product known as Version II (or Bluebook) Ethernet.

Subsequent to ratification this system has been known as IEEE 802.3. IEEE 802.3 is virtually, but not absolutely, identical to Bluebook Ethernet, in that they differ in two bytes within the frame.

There were many variants within the 802.3 standard. The better known ones are 10Base5, 10Base2, 10BaseT and 10BaseFL.

- 10Base5 is a 10 Mbps baseband system (Manchester encoding) that employs a bus topology using RG-8 coaxial cable with a maximum length of 500 m and 100 nodes maximum.
- 10Base2 is similar, but uses RG-58 coaxial cable. The maximum length is only 185 m and 30 nodes maximum can be attached to the segment.
- 10BaseT uses a hub (star) configuration. Nodes are connected to the hub with Cat 3 unshielded twisted-pair cable, with maximum length 100 m.
- 10BaseFL is a 2 km fiber-optic inter repeater link. It is used between two repeaters in order to connect two remote Ethernet segments and cannot have any nodes attached to it.

Subsequently, two additional specifications have been approved viz. IEEE 802.34 (100 Mbps or 'Fast' Ethernet) and IEEE 802.3z (1000 Mbps or 'Gigabit' Ethernet). Both fast and gigabit Ethernet have retained the original frame format, the only difference is that the frame is 10 or 100 times smaller, respectively. Fast and Gigabit Ethernet is also only available in a star configuration, using a hub.

IEEE 802.4 Token bus

The other major access method for a shared medium is the use of a token. This is a type of data frame that a station must possess before it can transmit messages. The stations are connected to a passive bus, although the token logically passes around in a cyclic manner.

This standard is the ratification of the token bus LAN developed by General Motors for its manufacturing automation protocol (MAP). The media used is usually broadband coax, and speeds vary from 1 to 10 Mbps.

IEEE 802.5 Token ring

As in 802.4, data transmission can only occur when a station holds a token. The logical structure of the network wiring is in the form of a ring, and each message must cycle through each station connected to the ring.

This standard is the ratified version of the IBM token ring LAN. However, where IBM token ring supports speed of 4 and 16 Mbps, IEEE 802.5 supports 1 and 4 Mbps. The physical media for the token ring can be either unshielded twisted pair, coaxial cable or optical fiber.

The original specification called for a single ring, which creates a problem if the ring gets broken. A subsequent enhancement of the specification, called IEEE 802.5u, introduce the concept of a dual redundant ring which enables the system to continue operating in case of a cable break.

IEEE 802.6 Metropolitan area networks

This committee is responsible for defining the standards for MANs. It has recommended that a system known as distributed queue data bus DQDB be utilized as a MAN standard.

The DQDB network is sponsored by Telecom Australia and defines the protocol for integrated voice and data on the same medium, within an area up to 15 km in diameter.

IEEE 802.7 Broadband LANs technical advisory group (TAG)

The 802.7 Committee provides technical advice on broadband technique.

IEEE 802.8 Fiber-optic LANs TAG

This is the fiber-optic equivalent of the 802.7 broadband TAG. The committee is attempting to standardize physical compatibility with FDDI and synchronous optical networks (SONET). It is also investigating single mode fiber and multimode fiber architectures.

IEEE 802.9 Integrated voice and data LANs

This committee has recently released a specification for isochronous Ethernet as IEEE 802.9a. It provides a 6.144 Mbps voice service (96 channels at 64 kbps) multiplexed with 10 Mbps data on a single cable. It is designed for multimedia applications.

IEEE 802.10 Secure LANs

Current proposals include two methods to address the lack of security in the original specifications. These are:

- A secure data exchange sublayer SDE sitting between the LLC and the MAC sublayer. There will be different SDEs for different systems i.e. military and medical.
- A secure interoperable LAN system architecture SILS. This will define system standards for secure LAN communications.

IEEE 802.11 Wireless LANs

The IEEE 802.11b standard WLANs (wireless LANs) operate in the 2.4 GHz band and feature a data rate of 11 Mbps. The IEEE has already issued a specification (IEEE 802.11a) for equipment operating at 5.7 GHz that supports a 54 Mbps data rate. It is expected that the 5.7 GHz band will eventually allow for a breakthrough to 100 Mbps.

IEEE 802.12 Fast LANs

This specification covers the system known as 100VG AnyLAN. Developed by Hewlett-Packard, this system operates on voice grade (CAT3) cable – hence the VG in the name. The AnyLAN indicates that the system can interface with both IEEE 802.3 and IEEE 802.5 networks (by means of a special speed adaptation bridge).

ANSI X3T9.5 FDDI

FDDI is a high sped token passing system that uses fiber optic as the medium. It is, in essence, a fast token ring system. It has the following characteristics:

- It supports speeds up to 100 Mbps
- It features dual contra-rotating rings for fault resilience
- It supports multiple tokens
- It supports up to 1000 nodes
- Nodes may be 2 km apart with multimode fiber, and 40 km with single-mode fiber.

IEEE 1394 Firewire

Firewire is a system originally designed to interconnect components on video cameras. It runs at a speed of 400 Mbps and supports of up to 63 devices on bus of up to 14 ft in length. It is hot-pluggable, i.e. devices can be connected or disconnected from the bus without affecting any participant on the bus. It also has power on the bus, so that devices do not have to be individually powered.

8.5   LAN extension and interconnecting devices

In the design of most LANs there are a number of different components that can be used such as repeaters, media converters, bridges, hubs, switches, routers and gateways. For simplicity this section will mainly deal with devices for Ethernet LANs.

The lengths of LAN segments are typically limited to a few hundred meters due to physical and collision domain constraints and there is often a need to increase this range. This can be achieved by means of a number of interconnecting devices, ranging from repeaters to gateways. It may also be necessary to partition an existing network for reasons of security or traffic overload.

In modern network devices the functions mentioned above are often mixed. Here are a few examples:

- A shared 10BaseT hub is, in fact, a multiport repeater.
- A level 2 switch is in essence a multiport bridge.
- Segmentable and dual-speed shared hubs make use of internal bridges.
- Level 2 switches can function as bridges, a two-port switch being none other than a bridge.
- Level 3 switches function as routers.

These examples are not meant to confuse the reader, but serve to emphasize the fact that the functions should be understood, rather than the 'boxes' in which they are packaged.

8.5.1   Repeaters

A repeater operates at the physical layer of the OSI model (level 1) and simply retransmits incoming electrical signals. This involves amplifying and retiming the signals received on one segment onto all other segments, without considering any possible collisions. All segments need to operate with the same media access mechanism and the repeater is unconcerned with the meaning of the individual bits in the packets. Collisions, truncated packets or electrical noise on one segment are transmitted onto all other segments (Figure 8.9).

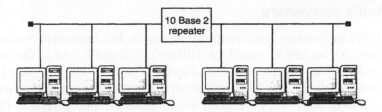

Figure 8.9
Repeater applications

Packaging

Repeaters are packaged either as standalone units (i.e. desktop models or small cigarette package-sized units) or 19 in. rack-mount units. Some of these can link two segments only, while larger rack-mount modular units (called concentrators) are used for linking multiple segments. Regardless of packaging, repeaters can be classified either as local repeaters (for linking network segments that are physically in close proximity), or as remote repeaters for linking segments that are some distance apart.

Local Ethernet repeaters

Several options are available:

- Two-port local repeaters offer most combinations of 10Base5, 10Base2, 10BaseT and 10BaseFL such as 10Base5/10Base5, 10Base2/10Base2, 10Base5/10Base2, 10Base2/10BaseT, 10BaseT/10BaseT and 10BaseFL/10BaseFL. By using such devices (often called boosters or extenders) it is possible, for example, to extend the distance between a computer and a 10BaseT hub by up to 100 m, or extend a 10BaseFL link between two devices (such as bridges) by up to 2 km.
- Multiport local repeaters offer several ports of the same type (e.g. 4 × 10Base2 or 8 × 10Base5) in one unit, often with one additional connector of a different type (e.g. 10Base2 for a 10Base5 repeater). In the case of 10BaseT the cheapest solution is to use an off-the-shelf 10BaseT shared hub, which is effectively a multiport repeater.
- Multiport local repeaters are also available as chassis-type units. i.e. as frames with common backplanes and removable units. An advantage of this approach is that 10Base2, 10Base5, 10BaseT and 10BaseFL can be mixed in one unit, with an option of SNMP management for the overall unit. These are also referred to as concentrators.

Remote repeaters

Remote repeaters, on the other hand, have to be used in pairs with one repeater connected to each network segment and a fiber-optic link between the repeaters. On the network side they typically offer 10Base5, 10Base2 and 10BaseT. On the interconnecting side the choices include 'single pair Ethernet', using telephone cable up to 457 m in length, or single mode/multimode optic fiber, with various connector options. With 10BaseFL (backwards compatible with the old FOIRL standard), this distance can be up to 1.6 km.

In conclusion it must be emphasized that although repeaters are probably the cheapest way to extend a network, they do so without in any way separating the collision domains, or network traffic. They simply extend the physical size of the network. All segments joined by repeaters therefore share the same bandwidth and collision domain.

Media converters

Media converters are essentially repeaters, but interconnect mixed media viz. copper and fiber. An example would be 10BaseT/10BaseFL. As in the case of repeaters, they are available in single and multiport options, and in stand-alone or chassis type configurations. The latter option often features remote management via SNMP.

Models may vary between manufacturers, but generally Ethernet media converters support:

- 10 Mbps (10Base2, 10BaseT, 10BaseFL – single and multimode)
- 100 Mbps (fast) Ethernet (100BaseTX, 100BaseFX – single and multimode) and
- 1000 Mbps (gigabit) Ethernet (single and multimode).

An added advantage of the fast and Gigabit Ethernet media converters is that they support full-duplex operation that effectively doubles the available bandwidth.

Switched repeaters

Switched repeaters (also known as packet repeaters) are significantly different to 'ordinary' repeaters. They do not simply pass frames across, but buffer each frame and check them for correctness. Each frame received is retransmitted following the CSMA/CD protocol, with the result that the repeater separates the two collision domains. For that reason there is no limitation on the number of repeaters that can be used.

8.5.2   Bridges

Bridges operate at the data link layer level of the OSI model (level 2) and are used to connect two separate networks to form a single large continuous LAN. The overall network, however, still remains one network with a single network ID (NetID). The bridge only divides the network up into two segments, each with its own collision domain and each retaining its full (e.g. 10 Mbps) bandwidth. Broadcast transmissions are seen by all nodes, on both sides of the bridge.

The bridge exists as an independent node on each network and passes only valid messages across to destination addresses on the other network. The decision as to whether or not a frame should be passed across the bridge is based on the level 2 address, i.e. the media (MAC) address. The bridge stores the frame from one network and examines its destination MAC address to determine whether it should be forwarded across the bridge.

Bridges can be classified as either MAC or LLC bridges, the MAC sublayer being the lower half of the data link layer and the LLC sublayer being the upper half. For MAC bridges the media access control mechanism on both sides must be identical; thus it can bridge only Ethernet to Ethernet, token ring to token ring and so on. For LLC bridges, the data link protocol must be identical on both sides of the bridge (e.g. IEEE 802.2 LLC); however the physical layers or MAC sublayers do not necessarily have to be the same. Thus the bridge isolates the media access mechanisms of the networks. Data can therefore be transferred, for example, between Ethernet and token ring LAN's. Collisions on the Ethernet system do not cross the bridge, nor do the tokens from the token ring.

Bridges can be used to extend the length of a network (as with repeaters) but in addition they improve network performance. If a network is demonstrating fairly slow response times, the nodes that mainly communicate with each other can be grouped together on one segment and the remaining nodes can be grouped together in another segment. The busy segment may not see much improvement in response rates (as it is already quite busy) but the lower activity segment may see quite an improvement in response times.

Bridges should be designed so that 80% or more of the traffic is within the LAN and only 20% crosses the bridge. Stations generating excessive traffic should be identified by a protocol analyzer and relocated to another LAN.

Intelligent bridges

Intelligent bridges (also referred to as transparent or spanning-tree bridges) are the most commonly used bridges because they are very efficient in operation and do not need to be taught the network topology. A transparent bridge learns and maintains two address lists corresponding to each network it is connected to. When a frame arrives from the one Ethernet network, its source address is added to the list of source addresses for that network. The destination address is then compared to that of the two lists of addresses for each network and a decision made whether to transmit the frame onto the other network. If no corresponding address to the destination node is recorded in either of these two lists, the message is retransmitted to all other bridge outputs (flooding), to ensure the message is delivered to the correct network. Over time, the bridge learns all the addresses on each network and thus avoids unnecessary traffic on the other network. The bridge also maintains time out data for each entry to ensure the table is kept up to date and old entries purged.

Transparent bridges cannot tolerate loops that could cause endless circulation of packets. If the network contains bridges that could form a loop as shown in Figure 8.10, one of the bridges (C) needs to be made redundant (de-activated).

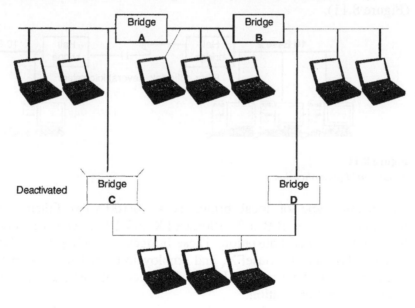

Figure 8.10
Avoidance of loops in bridge networks

The spanning tree algorithm (IEEE 802.1d) is used to manage paths between segments having redundant bridges. This algorithm designates one bridge in the spanning tree as the root and all other bridges transmit frames towards the root using a least cost metric. Redundant bridges can be re-activated if the network topology changes, such as when a circuit fails.

Source routing bridges

Source routing (SR) bridges are popular for IBM token ring networks. In these networks the sender must determine the best path to the destination. This is done by sending a discovery frame that circulates the network and arrives at the destination with a record of the path that was followed. These frames are returned to the sender who can then select the best path. Once the path has been discovered, the source updates its routing table and includes the path details in the routing information field in each transmitted frame.

SRT and translational bridges

When connecting Ethernet networks to token ring networks, either source routing transparent (SRT) bridges or translational bridges are used. SRT bridges are a combination of a transparent and source routing bridge, and are used to interconnect Ethernet (IEEE 802.3) and token ring (IEEE 802.5) networks. It uses source routing of the data frame if it contains routing information, otherwise it reverts to transparent bridging. Translational bridges, on the other hand, translate the routing information to allow source routing networks to bridge to transparent networks. The IBM 8209 is an example of this type of bridge.

Local vs remote bridges

Local bridges are devices that have two network ports and hence interconnect two adjacent networks at one point. Nowadays this function is usually performed by switches as they are basically multiport bridges, and are available with as few as two ports (Figure 8.11).

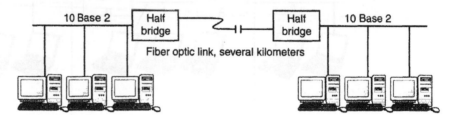

Figure 8.11
Remote bridge application

A useful type of local bridge is a 10/100 Mbps Ethernet bridge, which allows interconnection of 10BaseT, 100BaseTX and 100BaseFX networks, thereby performing the required speed translation. These bridges typically provide full-duplex operation on 100BaseTX and 100BaseFX, and employ internal buffers to prevent saturation of the 10BaseT port. They are built into dual-speed shared hubs in order to accomplish the required speed translation.

Remote bridges, on the other hand, operate in pairs with some form of interconnection between them. This interconnection can be with or without modems, and include RS-232/V.24, V.35, RS-422, RS-530, X.21, 4-wire, or fiber (both single and multimode). The distance between bridges can typically be up to 1.6 km.

8.5.3   Routers

Unlike bridges and level 2 switches, routers operate at layer 3 of the OSI model, namely at the network layer (or, the Internet layer of the DOD model). They therefore ignore address information contained within the data link level (the MAC addresses) and rather extract the address information contained in the network layer header. For TCP/IP, this is the IP address.

Like bridges or switches, routers appear as hosts on each network that they are connected to. They are connected to each participating network through a NIC, each with a MAC address as well as an IP address. Each NIC has to be assigned an IP address with the same NetID as the network it is connected to. This IP address allocated to each network is known as the default gateway for that network and each host on the internetwork requires at least one default gateway (but could have more). The default gateway is the IP address to which any host must forward a packet if it finds that the NetID of the destination and the local NetID does not match, which implies indirect delivery of the packet.

A second major difference between routers and bridges/switches is that routers will not act autonomously but rather have to be given the frames that need to be forwarded. A host has to forward a frame to the designated default gateway, else it will not be delivered.

Protocol dependency

Because routers operate at the network layer, they are used to transfer data between two networks that have the same Internet layer protocols (such as IP) but not necessarily the same physical or data link protocols. Routers are therefore said to be protocol dependent, and have to be able to handle all the Internet layer protocols present on a particular network. A network using Novell NetWare therefore requires routers that can accommodate IPX (Internet packet exchange) – the network layer component of SPX/IPX. If this network has to handle Internet access as well, it can only do this via IP, and hence the routers will need to be upgraded to models that can handle both IPX and IP.

Two-port vs multiport routers

Multiport routers are chassis-based devices with modular construction. They can interconnect several networks. The most common type of router is, however, a 2-port router. Since these are invariable used to implement WAN's, they connect LAN's to a 'communications cloud', the one port will be a local LAN port e.g. 10BaseT, but the second port will be a WAN port such as X.25.

Access routers

Access routers are 2-port routers that use dial-up access rather a permanent (e.g. X.25) connection to connect a LAN to an ISP and hence to the 'communications cloud' of the Internet. Typical options are ISDN or dial-up over telephone lines, using either the V.34 (ITU 33,6 kbps) or V.90 (ITU 56 kbps) standard. Some models allow multiple telephone lines to be used, using multilink PPP, and will automatically dial up a line when needed or redial when a line is dropped, thereby creating a 'virtual-leased line'.

Border routers

Routers within an autonomous system normally communicate with each other using an interior gateway protocol such as RIPv2. However, routers within an autonomous system that also communicate with remote autonomous systems need to do that via an exterior gateway protocol such as BGP-4. Whilst doing this, they still have to communicate with other routers within their own autonomous system, e.g. via RIP. These routers are referred to as border routers.

Routing vs bridging

It sometimes happens that a router is confronted with a level 3 (network layer) datagram header it cannot interpret. In the case of an IP router, this may be a Novell IPX address.

A similar situation will arise in the case of NetBIOS/NetBEUI, which is non-routable. A 'Brouter' (bridging router) will revert to a bridge if it cannot understand the level 3 protocol, and in this way forward the packet towards its destination. Most modern routers have this function built in.

8.5.4   Gateways

Gateways are network interconnection devices. They are not to be confused with default gateways (the IP addresses to which packets are forwarded for subsequent routing via indirect delivery).

A gateway is designed to connect dissimilar networks that cannot be connected at layers 2 or 3, and could operate anywhere from layer 4 of the OSI model or upwards. In a worst case scenario, a gateway may be required to decode and re-encode all seven layers of two dissimilar networks connected to either side, for example when connecting an Ethernet network to an IBM SNA network. Gateways thus have the highest overhead and the lowest performance of all the internetworking devices. The gateway translates from one protocol to the other and handles difference in physical signals, data format and speed.

Since gateways are, per definition, protocol converters, it so happens that a 2-port (WAN) router could also be classified as a gateway since it has to convert both layer 1 and layer 2 on the LAN side (say, Ethernet) to layer 1 and layer 2 on the WAN side (say, X.25). This leads to the confusing practice of referring to (WAN) IP routers as gateways.

8.5.5   Hubs

Hubs are used to implement physical star networks for Ethernet and token ring systems in such a way that electrical problems on individual node-to-hub links would not affect the entire network. This section will focus mainly on Ethernet hubs. Token ring hubs (multistation access units) are dealt with in Section 8.2.6.

Desktop vs stackable hubs

Smaller Ethernet desktop units are intended for stand-alone applications, and typically have five to eight ports. Some 10BaseT desktop models have an additional 10Base2 port. These devices are often called workgroup hubs.

Stackable hubs, on the other hand, typically have up to 24 ports and can be physically stacked and interconnected to act as one large hub without any repeater count restrictions. These stacks are often mounted in 19 in. cabinets.

Shared vs switched hubs

Shared hubs interconnect all ports on the hub in order to form a logical bus. This is typical of the cheaper workgroup hubs. All hosts connected to the hub share the available bandwidth since they all form part of the same collision domain.

Although they physically look alike, switched hubs (better known as switches) allow each port to retain and share its full bandwidth only with the hosts connected to that port. Each port (and the segment connected to that port) functions as a separate collision domain. This will be discussed in more detail, in the section on switches.

Managed hubs

A managed hub has an on-board processor with its own MAC and IP address. Once the hub has been set up via a PC on the hub's serial (COM) port, it can be monitored and controlled via the network using SNMP or RMON. The user can perform activities such as enabling/disabling individual ports, performing segmentation (see next section), monitoring the traffic on a given port, or setting alarm conditions for a given port.

Segmentable hubs

On a non-segmentable (i.e. shared) hub, all hosts share the same bandwidth. On a segmentable hub, however, the ports can be grouped, under software control, into several shared groups. All hosts on each segment then share the full bandwidth on that segment, which means that a 24 port 10BaseT hub segmented into four groups effectively supports 40 Mbps. The configured segments are internally connected via bridges, so that all ports can still communicate with each other if needed.

Dual-speed hubs

Some hubs offer dual-speed ports, e.g. 10BaseT/100BaseT. These ports are auto-configured, i.e. each port senses the speed of the NIC connected to it, and adjusts its own speed accordingly. All the 10BaseT ports connect to a common low speed internal segment, while all the 100BaseT ports connect to a common high speed internal segment. The two internal segments are interconnected via a speed-matching bridge.

Modular/chassis hubs

Some stackable hubs are modular, allowing the user to configure the hub by plugging in separate module for each port. Ethernet options typically include both 10 and 100 Mbps, with either copper or fiber. These hubs are sometimes referred to as chassis hubs.

Hub interconnection

Stackable hubs are best interconnected by means of special stacking cables attached to the appropriate connectors on the back of the chassis.

An alternative method for non-stackable hubs is by 'daisy-chaining' an interconnecting port on each hub by means of a UTP patch cord. Care has to be taken not to connect the transmit pins on the ports together (and, for that matter, the receive pins) – it simply will not work. A similar situation arises when interconnecting two COM ports with a 'straight' cable i.e. without a null modem. Connect transmit to receive and vice versa by (a) using a crossover cable and interconnecting two 'normal' ports, or (b) using a normal ('straight') cable and utilizing a crossover port on one of the hubs. Some hubs have a dedicated uplink (crossover) port while others have a port that can be manually switched into crossover mode.

A third method that can be used on hubs with a 10Base2 port is to create a backbone. This is done by attaching a BNC T-piece to each hub, and interconnecting the T-pieces with RG-58 co-axial cables. The open connections on the extreme ends of the backbone obviously have to be terminated.

Fast Ethernet hubs need to be deployed with caution because the inherent propagation delay of the hub is significant in terms of the 5.12 μs collision domain size. Fast Ethernet hubs are classified as Class I, II or II+, and the class dictates the number of hubs that can be interconnected. For example, ClassII dictates that there may be no more than two hubs between any given pair of nodes, that the maximum distance between the two hubs shall not exceed 5 m, and that the maximum distance between any two nodes shall not exceed 205 m. The safest approach, however, is to follow the guidelines of each manufacturer.

8.5.6   Switches

Ethernet switching is an expansion of the concept of bridging and switches are, in fact, multiport bridges. They enable frame transfers to be accomplished between any pair of devices on a network, on a per-frame basis. Only the two ports involved 'see' the specific frame. Illustrated in Figure (8.12) is an example of an 8-port switch, with 8 hosts attached. This comprises a physical star configuration, but it does not operate as a logical bus as an ordinary hub does. Since each port on the switch represents a separate segment with its own collision domain, it means that there are only two devices on each segment, namely the host and the switch port. Hence is no inherent risk of collisions!

In the sketch below hosts 1 and 7, 3 and 5, and 4 and 8 need to communicate at a given moment, and are connected directly for the duration of each frame transfer. For example, host 7 sends a packet to the switch, which determines the destination address, and directs the package to port 1 at 10 Mbps.

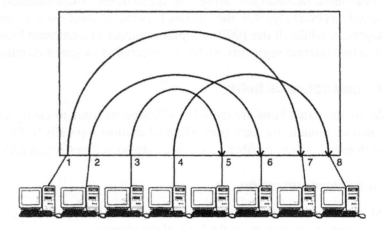

Figure 8.12
Ethernet switch

If host 3 wishes to communicate with host 5, the same procedure is repeated. Provided that there are no conflicting destinations, a 16-port switch could allow eight concurrent frame exchanges at 10 Mbps, rendering an effective bandwidth of 80 Mbps. On top of this, the switch could allow full-duplex operation which would double this figure.

Cut-through vs store-and-forward

Switches have two basic architectures, cut-through and store-and-forward. In the past, cut-through switches were faster because they examined the packet destination address only before forwarding the frame to the destination segment. A store-and-forward switch, on the other hand, accepts and analyses the entire packet before forwarding it to its destination. It takes more time to examine the entire packet, but it allows the switch to catch certain packet errors and keep them from propagating through the network. The speed of modern store-and-forward switches has caught up with cut-through switches so that the speed difference between the two is minimal. There are also a number of hybrid designs that mix the two architectures.

Since a store-and-forward switch buffers the frame, it can delay forwarding the frame if there is traffic on the destination segment, thereby adhering to the CSMA/CD protocol.

In the case of a cut-through switch this is a problem, since a busy destination segment means that the frame cannot be forwarded, yet it cannot be stored either. The solution is to force a collision back on the source segment, thereby forcing the source host to retransmit the frame.

Level 2 switches vs Level 3 switches

Level 2 switches operate at the data link layer of the OSI model and derive their addressing information from the destination MAC address in the Ethernet header. Level 3 switches, on the other hand, obtain addressing information from the network layer, e.g. from the destination IP address in the IP header.

Level 3 switches are used to replace routers in LANs as they can do basic IP routing (supporting protocols such as RIP and RIPv2) at almost 'wire-speed', hence they are significantly faster than routers.

Full-duplex switches

An additional advancement is full-duplex (fast) Ethernet where a device can simultaneously transmit AND receive data over one Ethernet connection. This requires a different Ethernet NIC in the host, as well as a switch that supports full-duplex. This enables two devices to exchange frames simultaneously via a switch. The host automatically negotiates with the switch and uses full-duplex if both devices support it.

Full-duplex is useful in situations where large amounts of data are to be moved around quickly, for example between graphics workstations and file servers.

Switch applications

Switches are very efficient in providing a high-speed aggregated connection to a server (Figure 8.13) or backbone. Apart from the normal lower speed (say, 10BaseT) ports, switches have a high speed uplink port (100BaseTX). This port is simply another port on the switch, accessible by all the other ports, but features a speed conversion from 10 to 100 Mbps.

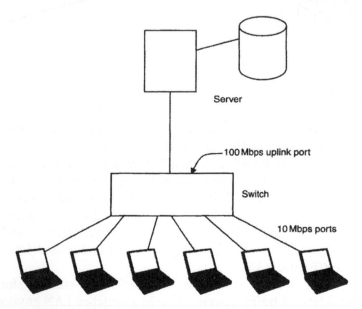

Figure 8.13
Switched connection to server

Assume that the uplink port was connected to a file server. If all the other ports (say, eight times 10BaseT) wanted to access the server concurrently, this would necessitate a bandwidth of 80 Mbps in order to avoid a bottleneck and subsequent delays. With a 10BaseT uplink port this would create a serious problem. However, with a 100BaseTX uplink there is still 20 Mbps of bandwidth to spare.

Switches are also very effective in backbone applications, linking several LANs together as one, yet segregating the collision domains. An example could be a switch located in the basement of a building, linking the networks on different floors of the building. Since the actual 'backbone' is contained within the switch, it is known in this application as a 'collapsed backbone' (Figure 8.14).

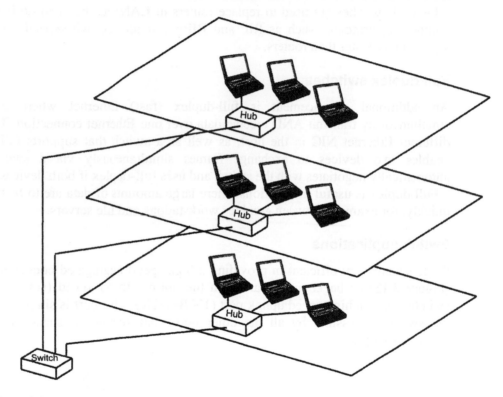

Figure 8.14
Using a switch as a backbone

8.6 VLANs

Provided that a Ethernet LAN is constructed around switches that support VLANs, individual hosts on the physical LAN can be grouped into smaller virtual LANs (VLANs), totally invisible to their fellow hosts. It is also possible to share resources such as servers or networkable printers between more than one VLAN. Unfortunately the 'standard' Ethernet/IEEE 802.3 header does not contain sufficient information to identify members of each VLAN, hence the frame had to be modified by the insertion of a 'tag', between the source MAC address and the type/length fields. This modified frame is known as an Ethernet 802.1Q tagged frame and is used for communication between the switches (Figure 8.15).

The IEEE 802.1p committee has defined a standard for packet-based LANs that supports layer 2 traffic prioritization in a switched LAN environment. IEEE 802.1p is part of a larger initiative (IEEE 802.1p/Q) that adds more information to the Ethernet header to allow networks to support VLANs and traffic prioritization.

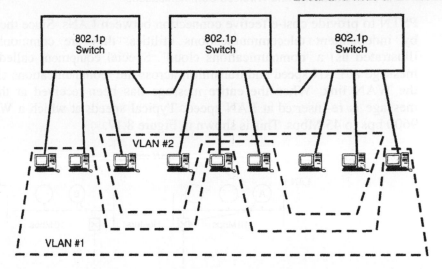

Figure 8.15
VLAN concept

8.7   MANs

An intermediate type of a network – a metropolitan area network – operates at speeds ranging from 56 kbps to 100 Mbps – typically a higher speed than a WAN but slower than a LAN. MANs use fiber-optic technology to communicate over distances of up to several hundred kilometers and use a ring topology. They are normally used by utilities or telecommunication service providers within cities (Figure 8.16).

Network technologies applicable here include IEEE 802.6 DQDB and ANSI X3T9.5 FDDI.

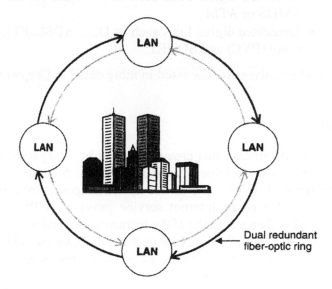

Figure 8.16
MAN concept

8.8   WANs

While LANs operate where distances are relatively small, wide area networks (WANs) are used to link LANs that are separated by large distances that range from a few tens of meters to thousands of kilometers. WANs normally use the services supplied by the

PSTN to provide cost-effective connection between LANs. Since these links are supplied by independent telecommunications utilities, they are commonly referred to (and illustrated as) a 'communications cloud'. Special equipment called gateways store the message at LAN speed and transmit it across the 'communications cloud' at the speed of the WAN link. When the entire message has been received at the remote LAN, the message is re-inserted at LAN speed. Typical speeds at which a WAN interconnects is 9600 bps to 45 Mbps. This is shown in Figure 8.17.

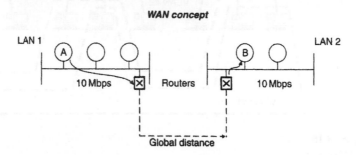

Figure 8.17
WAN concept

The communications cloud encompasses the various WAN transport alternatives. These alternatives include:

- Switched analog lines
- Dedicated analog lines, either unconditioned or conditioned C or D type lines
- Switched digital lines such as switched 56, ISDN (BRI) frame relay (SVC), SMDS or ATM
- Dedicated digital lines such as DDS, xDSL, FT1, T-1, T-3, ISDN (PRI), frame relay (PVC) and SONET.

These alternatives are discussed in more detail in Chapter 6.

8.9   VPNs

Businesses are finding that creating WANs by using dedicated leased line or frame-relay circuits are relatively expensive and inflexible. A cheaper alternative is to create a virtual private network (VPN) and to set up a connection from each LAN to a local point of presence (POP) of an Internet service provider (ISP). The details of getting the data transported is then left to the ISP's infrastructure and the Internet. VPNs are not limited to corporate sites, but allow mobile users to dial into the VPN via the POP nearest to them, eliminating the need for banks of modems and remote access servers (RAS) at the corporate site (Figure 8.18).

A potential problem is the fact that the traffic between the networks share all the other Internet traffic and hence the VPN is theoretically accessible to everyone on the Internet. VPNs therefore need to provide four functions to ensure security for all users. They are:

1. Authentication, i.e. ensuring that the data has actually originated from the source that it claims.
2. Access control, i.e. only allowing legitimate users on to the network.

3. Data integrity, i.e. ensuring that nobody can tamper with the data whilst traveling on the Internet.
4. Confidentiality, i.e. ensuring that nobody can view or copy data as it travels across the Internet.

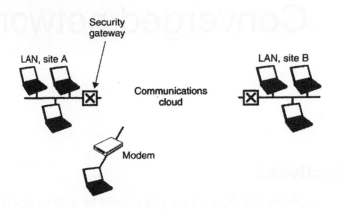

Figure 8.18
VPN concept

The authentication problem is addressed by various password-based systems and challenge – response systems such as challenge handshake authentication protocol (CHAP) and remote authentication dial-in user service (RADIUS). Hardware-based tokens and digital certificates used to control access to network resources.

The integrity and confidentiality problem arises since the packets are visible as they travel across the Internet. This is taken care of by tunneling and encryption. A number of protocols have been developed in order to create tunnels, allowing users to encapsulate their communication in IP packets. Two types of tunnels are normally used, namely LAN-to-LAN tunnels existing between the security gateway (router or firewall) at each LAN, or a client-to-LAN tunnel between a mobile user and the security gateway at a LAN.

Four different protocols have been proposed for tunneling across the Internet. They are point-to-point tunneling protocol (PPTP), Layer-2-forwarding (L2F), Layer-2-tunneling protocol (L2TP) and IP security protocol (IPSec). PPTP, L2F and L2TP are aimed at dial-up VPNs while IPSec is aimed at LAN-to-LAN communications.

IPSec is generally considered to best solution since it includes strong security measures with regard to encryption, authentication and key management. IPSec allows the sender, or the router/firewall acting on his behalf, to authenticate, encrypt, or apply both operations on each individual IP packet. IPSec also operates in two modes. In transport mode, it authenticates or encrypts only the transport-layer segment of the IP packet. In tunnel mode, it applies the authentication or encryption to the entire packet. The encryption methods involve:

- Diffie-Hellman key exchanges for delivering private (secret) keys between two parties
- Public-key cryptography for signing Diffie-Hellman exchanges to guarantee the identity of the two parties
- Data encryption standard (DES) for encrypting data (bulk encryption)
- Keyed hash algorithms (HMAC, MD5, SHA) for authenticating packets
- Digital certificates for validating public keys.

IPSec is, unfortunately, designed for IP only and therefore PPTP and L2TP are good alternatives for non-IP environments e.g., those using AppleTalk, IPX and NetBEUI.

9

Converged networks

Objectives

After completion of this chapter you should be able to describe, in basic terms:

- The TCP/IP packet transport mechanism, with particular reference to IPv4, IPv6 and ICMP
- The TCP/IP routing mechanism, with particular reference to the interior gateway protocols RIP and OSPF, and the exterior gateway protocol BGP-4
- The TCP/IP end-to-end protocols TCP and UDP
- The protocols supporting VoIP, in particular multicast IP, RTP, RTCP, RSVP and RTSP
- The H.323 standard, with reference to

  - The H.323 hardware, i.e. terminals, gateways, gatekeepers and MCUs
  - The H.323 protocols, i.e. H.225 and H.245
  - Audio and video codecs.

9.1 Introduction

This chapter deals with the convergence of conventional PSTN networks and IP-based internetworks, in particular the Internet. As a result of this convergence, voice over IP (VoIP) is making major inroads into the telecommunications industry. This chapter will introduce the ITU-T H.232 standard for multimedia (audio, video and data) transmission.

H.323 requires IP as a network layer protocol and TCP/UDP as transport layer protocols. The first part of the chapter will therefore deal with the TCP/IP protocol suite, as a necessary and sufficient basis for the understanding of the operation of H.323 and the concept of VoIP.

The second part of this chapter will deal with the H.323 standard itself.

9.2 Applications

The major advantage of VoIP over conventional PSTN usage is cost. It is not uncommon for a large company to recover the installation cost of VoIP equipment in less than a year. The following are a couple of implementations.

Interoffice networking Instead of having separate voice and data networks within the building, employees can use both their telephones and computers on one line.

Outbound calls By using the company's existing infrastructure (X.25, ATM, frame relay, etc.) for outbound calls, the company can save on-line rental by reducing the number of outgoing trunk lines.

Internet surfing at home Using, for example, itRings in the USA and Telstra iRing in Australia, users with only one line at home can answer incoming calls on their PCs while surfing the Internet.

Long distance outbound calls Companies making a large percentage of their long-distance outbound calls to a few area codes (typical of countries like South Africa and Australia) can install H.323 Gateways in those areas, with the result that for billing purposes these calls become local calls.

Inbound customer (1–800) charges The same philosophy applies. By placing H.323 gateways in areas where most inbound calls originate, those inbound calls appear as local calls to the callers. The only cost to the vendor is the gateway installation.

9.3 Protocols

9.3.1 Introduction

TCP/IP is the de facto global standard for the network and transport layer implementation of internetwork applications because of the popularity of the Internet. The Internet (in its early years known as ARPANet), was part of a military project commissioned by the Advanced Research Projects Agency (ARPA), later known as the Defence Advanced Research Agency or DARPA. The communications model used to construct the system is known as the ARPA model.

Whereas the OSI model was developed in Europe by the International Standards Organisation (ISO), the ARPA model (also known as the DoD model) was developed in the USA by ARPA. Although they were developed by different bodies and at different points in time, both serve as models for a communications infrastructure and hence provide 'abstractions' of the same reality. The remarkable degree of similarity is therefore not surprising.

Whereas the OSI model has seven layers, the ARPA model has four layers. The OSI layers map onto the ARPA model as follows:

- The OSI session, presentation and applications layers are contained in the ARPA process and application layer.
- The OSI transport layer maps onto the ARPA host-to-host layer (sometimes referred to as the service layer).
- The OSI network layer maps onto the ARPA Internet layer.
- The OSI physical and data link layers map onto the ARPA network interface layer.

The relationship between the two models is depicted in Figure 9.1.

TCP/IP, or rather – the TCP/IP Protocol Suite – is not limited to the TCP and IP protocols, but consist of a multitude of interrelated protocols that occupy the upper three layers of the ARPA model. TCP/IP does NOT include the bottom network interface layer, but depends on it for access to the medium.

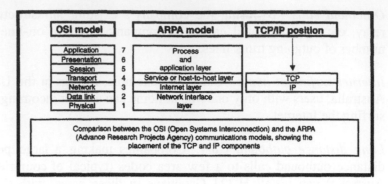

Figure 9.1
OSI vs ARPA models

As depicted in Figure 9.2, an Internet transmission frame originating on a specific host (computer) would contain the local network (e.g. Ethernet) header and trailer applicable to that host. As the message proceeds along the Internet, this header and trailer could be replaced depending on the type of network on which the packet finds itself – be that X.25, frame relay or ATM. The IP datagram itself would remain untouched, unless it has to be fragmented and re-assembled along the way.

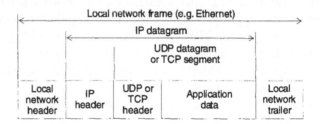

Figure 9.2
Internet frame

(*Note*: Any Internet-related specification is referenced as a request for comments or RFC. RFCs can be obtained from various sources on the Internet such as www.rfc-editor.org.)

The Internet layer

This layer is primarily responsible for the routing of packets from one host to another. Each packet contains the address information needed for its routing through the internetwork to the destination host. The dominant protocol at this level is the Internet protocol (IP). There are, however, several other additional protocols required at this level such as:

- *Address resolution protocol (ARP), RFC 826*:   This is used for the translation of an IP address to a hardware (MAC) address, such as required by Ethernet.
- *Reverse address resolution protocol (RARP), RFC 903*:   This is the complement of ARP and translates a hardware address to an IP address.
- *Internet control message protocol (ICMP), RFC 792*:   This is a protocol used for exchanging control or error messages between routers or hosts.

The host-to-host layer

This layer is primarily responsible for data integrity between the sender host and receiver host regardless of the path or distance used to convey the message. It has two protocols associated with it, namely:

1. User data protocol (UDP), a connectionless (unreliable) protocol used for higher layer port addressing with minimal protocol overhead (RFC 768).
2. Transmission control protocol (TCP), a connection-oriented protocol that offers a very reliable method of transferring a stream of data in byte format between applications (RFC 793).

The process and application layer

This layer provides the user or application programs with interfaces to the TCP/IP stack. Some application layer protocols relevant to this chapter include:

- Real time protocol (RTP), which provides end-to-end delivery services for data that requires real-time support, such as voice and data (RFC 1889).
- Real time control protocol (RTCP), a companion protocol for RTP, which monitors the quality of service and conveys information about participants during an RTP session (RFC 1889).
- Resource reservation protocol (RSVP), which reserves appropriate levels of service from all devices interconnecting two hosts sharing data that need a high level of quality of service (RFC 2205).
- Real time streaming protocol (RTSP), which remotely controls the delivery of data with real-time properties, such as audio and video (RFC 2326).
- Session description protocol (SDP), which describes multimedia sessions for the purpose of announcement and invitation (RFC 2327).

Other protocols at this level include (but are not limited to) file transfer protocol (FTP), trivial file transfer protocol (TFTP), simple mail transfer protocol (SMTP), telecommunications network (TELNET), post office protocol (POP3), remote procedure calls (RPC), remote login (RLOGIN), hypertext transfer protocol (HTTP) and network time protocol (NTP). Users can also develop their own application layer protocols by means of a developer's kit such as Winsock.

9.3.2 Internet layer protocols (packet transport)

This section will deal with the Internet protocol (IP) and the Internet control message protocol (ICMP).

IP version 4 (IPv4)

IP (RFC 791) is responsible for the delivery of packets ('datagrams') between hosts. It is analogous to the postal system, in that it forwards (routes) and delivers datagrams on the basis of IP Addresses attached to the datagrams, in the same way the postal service would process a letter based on the postal address. The IP Address is a 32-bit entity containing both the network address (the 'zip code') and the host address (the 'street address').

IP also breaks up (fragments) datagrams that are too large. This is often necessary because the LANs and WANs that a datagram may have to traverse on its way to its destination may have different frame size limitations. For example, Ethernet can handle 1500 bytes but X.25 can handle only 576 bytes. IP on the sending side will fragment a datagram if necessary, attach an IP header to each fragment, and send them off consecutively. On the receiving side, IP will again rebuild the original datagram.

The IPv4 header

The IP header is appended to the information that IP accepts from higher-level protocols, before passing it around the network. This information could, within itself, contain the headers appended by higher level protocols such as TCP. The header consists of at least five 32-bit (4 byte) 'long words' i.e. 20 bytes total and is made up as in Figure 9.3.

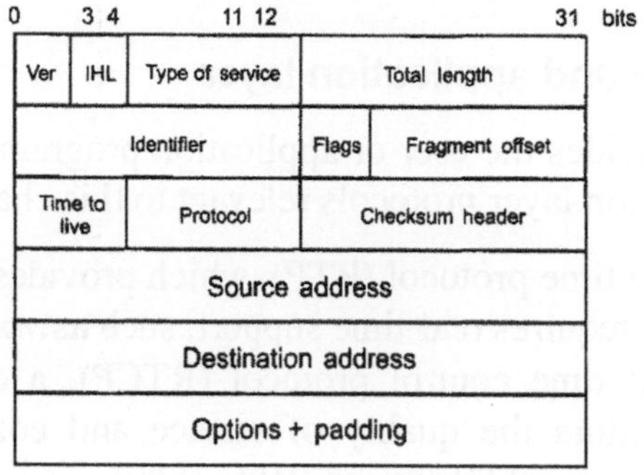

Figure 9.3
IPv4 header

The Ver (version) field is 4 bits long and indicates the version of the IP protocol in use. For IPv4 it is 4. This is followed by the 4-bit IHL (Internet header length) field that indicates the length of the IP header in 32-bit 'long words'. This is necessary since the IP header can contain options and therefore does not have a fixed length.

The 8-bit type of service (ToS) field informs the network about the quality of service required for this datagram. The ToS field is composed of a 3-bit precedence field (which is often ignored) and an unused (LSB) bit that must be 0. The remaining 4 bits may only be turned on (set = 1) one at a time, and are allocated as follows:

Bit 3: Minimize delay
Bit 4: Maximize throughput
Bit 5: Maximize reliability
Bit 6: Minimize monetary cost.

Total length (16 bits) is the length of the entire datagram, measured in bytes. Using this field and the IHL length, it can be determined where the data starts and ends. This field allows the length of a datagram to be up to $2^{16} = 65\ 536$ bytes, although such long datagrams are impractical. All hosts must at least be prepared to accept datagrams of up to 576 octets.

The 16-bit identifier uniquely identifies each datagram sent by a host. It is normally incremented by one for each successive datagram sent. In the case of fragmentation, it is appended to all fragments of the same datagram for the sake of reconstructing the datagram at the receiving end. It can be compared to the 'tracking' number of an item delivered by registered mail or UPS.

The 3-bit flag field contains two flags, used in the fragmentation process, viz. DF and MF. The df (don't fragment) flag is set (=1) by the higher level protocol (e.g. TCP) if IP is NOT allowed to fragment a datagram. If such a situation occurs, IP will not fragment and forward the datagram, but simply return an appropriate ICMP error message to the sending host. If fragmentation does occur, MF = 1 will indicate that there are more fragments to follow, whilst MF = 0 indicates that it is the last fragment to be sent.

The 13-bit fragment offset field indicates where in the original datagram a particular fragment belongs, i.e. how far the beginning of the fragment is removed from the end of the header. The first fragment has offset zero. The fragment offset is measured in units of 8 bytes (64 bits); i.e. the transmitted offset is equal to the actual offset divided by eight.

The TTL (time to live) field ensures that undeliverable datagrams are eventually discarded. Every router that processes a datagram must decrease the TTL by one and if this field contains the value zero, then the datagram must be destroyed. Typically a datagram can be delivered anywhere in the world by traversing fewer than 15 routers.

The 8-bit protocol field indicates the next (higher) level protocol header present in the data portion of the IP datagram, in other words the protocol that resides above IP in the protocol stack and which has passed the datagram down to IP. Typical values are 1 for ICMP, 6 for TCP and 17 for UDP. A more detailed listing is contained in RFC1700.

The checksum is a 16-bit mathematical checksum on the header only. Since some header fields change at each hop (e.g. TTL), this checksum is recomputed and verified at each point that the IP header is processed. It is not necessary to cover the data portion of the datagram, as the protocols making use of IP, such as ICMP, IGMP, UDP and TCP, all have a checksum in their headers to cover their own header and data.

Finally, the source and destination addresses are the 32-bit IP addresses of the origin and the destination hosts of the Datagram.

IPv4 addressing

The ultimate responsibility for the issuing of IP addresses is vested in the Internet assigned numbers authority (IANA). This responsibility is, in turn, delegated to the three regional Internet registries (RIRs) viz. APNIC (Asia-Pacific Network Information Center), ARIN (American Registry for Internet Numbers) and RIPE NCC (Reseau IP Europeans). RIRs allocate blocks of IP addresses to Internet service providers (ISPs) under their jurisdiction, for subsequent issuing to users or subISP's.

The IPv4 address consists of 32 bits, e.g. 11000000011001000011001 0000000001. Since this number is fine for computers but a little difficult for human beings, it is divided into four octets w, x, y and z. Each octet is converted to its decimal equivalent. The result of the conversion is written in the format 192.100.100.1. This is known as the 'dotted decimal' or 'dotted quad' notation. As mentioned earlier, one part of the IP address is known as the network ID or 'NetID' while the rest is known as the 'HostID' (Figure 9.4).

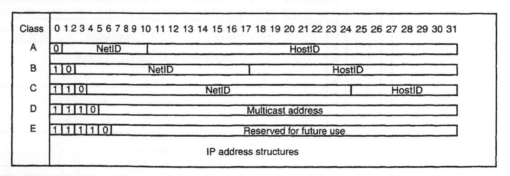

Figure 9.4
Address structure for IPv4

Originally, IP addresses were allocated in so-called address classes. Although the system proved to be problematic, and IP addresses are currently issued 'classless', the legacy of IP address classes remains and has to be understood.

To provide for flexibility in assigning addresses to networks, the interpretation of the address field was coded to specify either a small number of networks with a large number of hosts (Class A), or a moderate number of networks with a moderate number of hosts (Class B), or a large number of networks with a small number of hosts (Class C). There was also provision for extended addressing modes: Class D was intended for multicasting whilst E was reserved for future use.

For Class A, the first bit is fixed at 0. The values for 'w' can therefore only vary between 0 and 12 710. The value 0 is not allowed and 127 is a reserved number used for testing. This allows for 126 Class A NetIDs. The number of HostIDs is determined by octets 'x', 'y' and 'z'. From these 24 bits, $2^{24} = 16\ 777\ 218$ combinations are available. All 0s and all 1s are not permissible, which leaves 16 777 216 usable combinations.

For Class B, the first 2 bits are fixed at 10. The binary values for 'w' can therefore only vary between 12 810 and 19 110. The number of NetIDs is determined by octets 'w' and 'x'. The first 2 bits are used to indicate class B and hence cannot be used. This leaves fourteen usable bits. Fourteen bits allow $2^{14} = 16\ 384$ NetIDs. The number of HostIDs is determined by octets 'y' and 'z'. From these 16 bits, $2^{16} = 65\ 536$ combinations are available. All 0s and all 1s are not permissible, which leaves 65 534 usable combinations.

For Class C, the first 3 bits are fixed at 110. The binary values for 'w' can therefore only vary between 19 210 and 22 310. The number of NetIDs is determined by octets 'w', 'x' and 'y'. The first 3 bits (110) are used to indicate class C and hence cannot be used. This leaves 22 usable bits. Twenty-two bits allow $2^{22} = 2\ 097\ 152$ combinations for NetIDs. The number of HostIDs is determined by octet 'z'. From these 8 bits, $2^{8} = 256$ combinations are available. Once again, all 0s and all 1s are not permissible which leaves 254 usable combinations.

In order to determine where the NetID ends and the HostID begins, each IP address is associated with a sub-net mask, or, technically more correct, a netmask. This mask starts with a row of contiguous 1s from the left; one for each bit that forms part of the NetID. This is followed by 0s, one for each bit comprising the HostID.

IP version 6 (IPv6)

IPv4 has several shortcomings, the major one being the 32-bit IP address that (only!) allows for 2^{32} different IP addresses. As early as 1993 it was realized that the world was running out of IP addresses and the Internet Engineering Task Force (IETF) chartered the Internet Protocol Next Generation (IPng) working group (Figure 9.5).

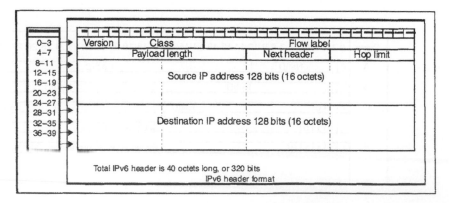

Figure 9.5
IPv6 header

IPv6 (IPng) differs from IPv4 in several respects.

- The 128-bit network address allows for 2^{96} times more IP addresses – in reality several thousand IP addresses for each square meter of the earth!
- The header structure is more efficient. Despite the basic header is only twice the length of the IPv4 header.
- IPv6 allows extension headers to be added when required for additional applications and options.
- There is no header checksum.
- A flow label allows for quality of service requirements.
- It has built-in security for authentication and encryption.

Here follows a brief description of the header fields:

- The 4-bit version number in the IP header holds the IP version number (6 for IPv6).
- The 8-bit traffic class field holds a value indicating the datagram's delivery criteria.
- The flow label field is 24 bits in length and a host may use it to specify special handling for certain packages.
- The 16-bit payload length field contains an unsigned integer to specify the total length of the IP datagram in bytes. Payloads bigger than 65 535 are allowed, and are called Jumbo payloads.
- The next header field, similar to the protocol field in IPv4, is 1 byte in length and identifies the header immediately following the IPv6 header. This is the same as the IPv4 protocol field. Some examples are shown in the table below: If there is no extension header, it points to the protocol 'above' IP e.g. TCP or UDP.

Destination options

The hop limit field, similar to the TTL field in IPv4, is 1 byte in length determines the number of hops the datagram can travel. It is decremented at each node, and an error message sent back if it becomes zero.

The source and destination IP addresses in 128-bit format are placed in the header. The destination address is for the recipient of the package and may *not* be the ultimate recipient if a routing header is present.

Extension headers

In order to reduce the size of the header and to reduce the time needed to process it, options have been removed from the main header and are only appended when needed.

The hop-by-hop extension header, when added, is processed by all intermediate routers and therefore has to follow directly after the main header. The other extension headers are only processed at the final destination and are added individually, if needed, in the following sequence:

- IPv6 header
- Hop-by-hop options header
- Destination options header (for options processed by the first destination that appears in the Ipv6 destination address field, plus any subsequent destinations listed in the routing header)
- Routing header

- Fragment header
- Authentication header
- Encapsulating security payload header
- Destination options header (for options to be processed by the final destination only).

Figure 9.6 illustrates the IPv6 and optional headers, in their suggested order.

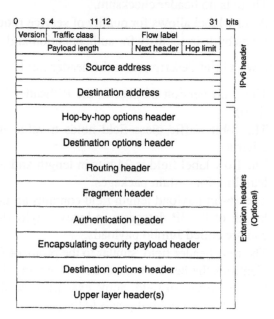

Figure 9.6
IPv6 extension headers

IPv6 addresses

IPv6 addresses are 128 bits long and are identifiers for individual interfaces or sets of interfaces. All IPv6 addresses are assigned to interfaces (i.e. network interface cards) and *not* to nodes i.e. hosts. Since each interface belongs to a single node, any of that node's interfaces' unicast addresses may be used as an identifier for the node. A single interface may be assigned multiple IPv6 addresses of any type.

There are three types of IPv6 addresses. These are unicast, anycast and multicast.

1. Unicast addresses identify a single interface.
2. Anycast addresses identify a set of interfaces such that a packet sent to an anycast address will be delivered to one member of the set.
3. Multicast addresses identify a group of interfaces, such that a packet sent to a multicast address is delivered to all of the interfaces in the group. There are no broadcast addresses in IPv6, their function being superseded by multicast addresses.

Addresses are written in the format x:x:x:x:x:x:x:x where each x represents 16 bits in hexadecimal format, for example 1432:FEDA:AABB:1234:5678:9ABC:1111:9999.

Four 0s (:0000: can be abbreviated as :0:. A series of leading and trailing 0s can be written as :: (double colon).

The leading bits in the address indicate the specific type of IPv6 address. The variable length field comprising these leading bits is called the format prefix (FP). The initial allocation of these prefixes is shown in Table 9.1.

Allocation	Prefix (Binary)	Allocated Fraction of Address Space
Reserved	0000 0000	1/256
Unassigned	0000 0001	1/256
Reserved for NSAP allocation	0000 001	1/128
Reserved for IPX allocation	0000 010	1/128
Unassigned	0000 011	1/128
Unassigned	0000 1	1/32
Unassigned	0001	1/16
Unassigned	001	1/8
Aggragatable global unicast address	010	1/8
Unassigned	011	1/8
Reserved for geographic-based		
Unicast addresses	100	1/8
Unassigned	101	1/8
Unassigned	110	1/8
Unassigned	1110	1/16
Unassigned	1111 0	1/32
Unassigned	1111 10	1/64
Unassigned	1111 110	1/128
Unassigned	1111 1110 0	1/512
Link local use addresses	1111 1110 10	1/1024
Site local use addresses	1111 1110 11	1/1024
Multicast addresses	1111 1111	1/256

Table 9.1
IPv6 address ranges

Unicast addresses
There are several forms of unicast address assignment in IPv6. These are:

- Aggragatable global unicast addresses
- Unspecified addresses
- Loopback addresses
- IPv4-based addresses
- Site local addresses
- Link local addresses.

Aggragatable global unicast addresses are used for global communication. The first 3 bits identify the address as an aggragatable global unicast address. The remainder of the bits is sub-divided into fields in order to provide a hierarchical identification of the network providers, networks, sub-networks and end-user devices.

Unspecified addresses can be written as 0:0:0:0:0:0:0:0, or simply ‘::’. These can be used as source addresses by stations that have not yet been configured with IP addresses. They can never be used as destination addresses. This is similar to 0.0.0.0 in IPv4.

Loopback addresses (0:0:0:0:0:0:0:1) can be used by a node to send a datagram to itself. It is similar to the 127.0.0.1 of IPv4.

An IPv4-based IPv6 address can be constructed out of an existing IPv4 address. This is done by prepending 96 zero bits to a 32-bit IPv4 address. The result is written as 0:0:0:0:0:0:192.100.100.3, or simply ::192.100.100.3.

Site local addresses are partially equivalent of the IPv4 private addresses (i.e. addresses for private use that cannot be used on the Internet). A typical site local address will consist of the relevant prefix, a set of 38 zeros, a sub-net ID, and the interface identifier. Site local addresses cannot be routed in the Internet, but only between two stations on a single site.

Link local addresses are used by stations that are not yet configured with either a provider-based address or a site local address may use link local addresses. Theses are composed of the link local prefix, 1111 1110 10, a set of 0s, and an interface identifier. These addresses can only be used by stations connected to the same local network and packets addressed in this way cannot traverse a router.

Internet control message protocol

When nodes fail, or become temporarily unavailable, or when certain routes become overloaded with traffic, a message mechanism called the Internet control message protocol (ICMP) reports errors and other useful information about the performance and operation of the network.

ICMP communicates between the Internet layers on two nodes and is used by routers as well as individual hosts. Although ICMP is viewed as residing within the Internet layer, its messages travel across the network encapsulated in IP datagrams in the same way as higher layer protocol (such as TCP or UDP) datagrams. The ICMP message, consisting of an ICMP header and ICMP data, is encapsulated as 'data' within an IP datagram that is, in turn, carried as 'payload' by the lower network interface layer (for example, Ethernet).

There are a variety of ICMP messages, each with a different format, yet the first three fields as contained in the first 4 bytes or 'long word' is the same for all.

The various ICMP messages are shown in Figure 9.7.

The three common fields are:

1. An ICMP message type (4 bits) which is a code that identifies the type of ICMP message.
2. A code (4 bits) in which interpretation depends on the type of ICMP message.
3. A checksum (16 bits) that is calculated on the entire ICMP datagram.

Table 9.2 lists the different types of ICMP messages.

ICMP Messages can be further sub-divided into two broad groups viz. ICMP error messages (destination unreachable, time exceeded, invalid parameters, source quench or redirect) and ICMP query messages (echo request and reply messages, timestamp request and reply messages, and sub-net mask request and reply messages).

Here follows a few examples of ICMP error messages.

Source quench

If a gateway (router) receives a high rate of datagrams from a particular source it will issue a source quench ICMP message for every datagram it discards. The source node will then slow down its rate of transmission until the source quench messages stop, at which stage it will gradually increase the rate again.

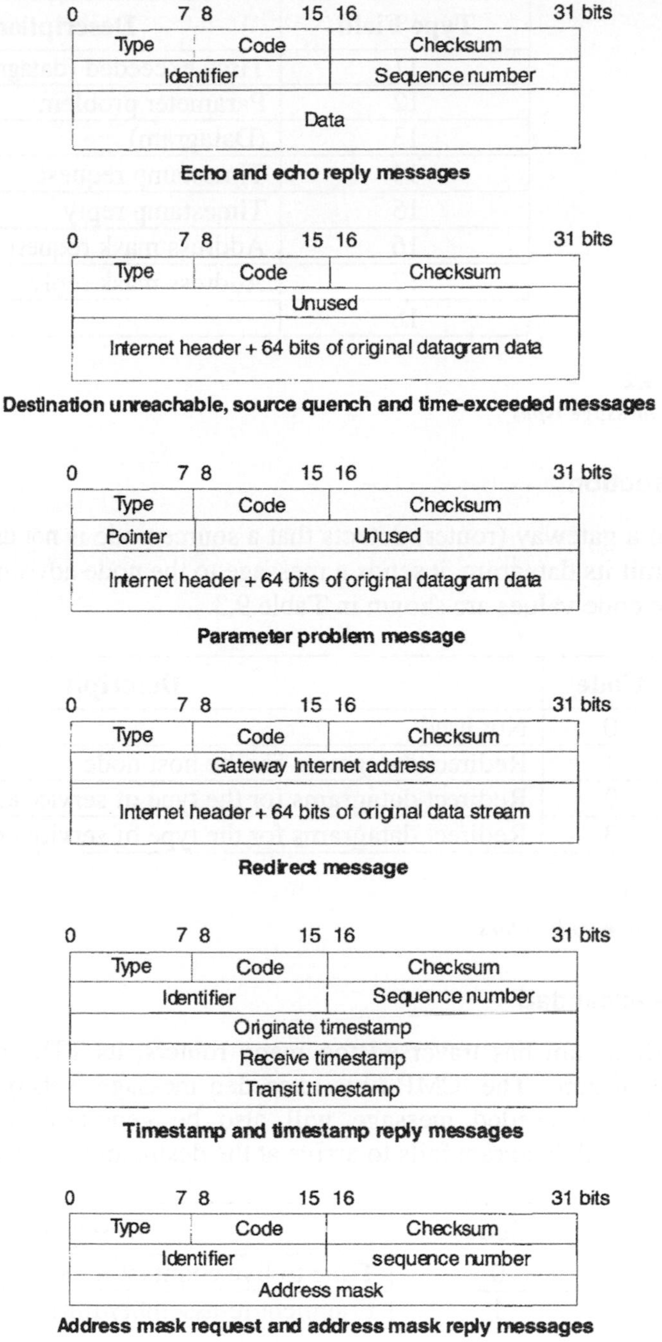

```
0              7 8            15 16                      31 bits
┌──────────────┬──────────────┬────────────────────────────┐
│     Type     │     Code     │         Checksum           │
├──────────────┴──────────────┼────────────────────────────┤
│          Identifier         │      Sequence number       │
├─────────────────────────────┴────────────────────────────┤
│                          Data                             │
└───────────────────────────────────────────────────────────┘
```
Echo and echo reply messages

```
0              7 8            15 16                      31 bits
┌──────────────┬──────────────┬────────────────────────────┐
│     Type     │     Code     │         Checksum           │
├──────────────┴──────────────┴────────────────────────────┤
│                        Unused                             │
├───────────────────────────────────────────────────────────┤
│     Internet header + 64 bits of original datagram data   │
└───────────────────────────────────────────────────────────┘
```
Destination unreachable, source quench and time-exceeded messages

```
0              7 8            15 16                      31 bits
┌──────────────┬──────────────┬────────────────────────────┐
│     Type     │     Code     │         Checksum           │
├──────────────┼──────────────┴────────────────────────────┤
│    Pointer   │                Unused                      │
├──────────────┴───────────────────────────────────────────┤
│     Internet header + 64 bits of original datagram data   │
└───────────────────────────────────────────────────────────┘
```
Parameter problem message

```
0              7 8            15 16                      31 bits
┌──────────────┬──────────────┬────────────────────────────┐
│     Type     │     Code     │         Checksum           │
├──────────────┴──────────────┴────────────────────────────┤
│                 Gateway Internet address                  │
├───────────────────────────────────────────────────────────┤
│    Internet header + 64 bits of original data stream      │
└───────────────────────────────────────────────────────────┘
```
Redirect message

```
0              7 8            15 16                      31 bits
┌──────────────┬──────────────┬────────────────────────────┐
│     Type     │     Code     │         Checksum           │
├──────────────┴──────────────┼────────────────────────────┤
│          Identifier         │      Sequence number       │
├─────────────────────────────┴────────────────────────────┤
│                  Originate timestamp                      │
├───────────────────────────────────────────────────────────┤
│                   Receive timestamp                       │
├───────────────────────────────────────────────────────────┤
│                   Transit timestamp                       │
└───────────────────────────────────────────────────────────┘
```
Timestamp and timestamp reply messages

```
0              7 8            15 16                      31 bits
┌──────────────┬──────────────┬────────────────────────────┐
│     Type     │     Code     │         Checksum           │
├──────────────┴──────────────┼────────────────────────────┤
│          Identifier         │      sequence number       │
├─────────────────────────────┴────────────────────────────┤
│                      Address mask                         │
└───────────────────────────────────────────────────────────┘
```
Address mask request and address mask reply messages

Figure 9.7
ICMP message formats

Type Field	Description
0	Echo, reply
3	Destination unreachable
4	Source quench
5	Redirect (change a route)
8	Echo request

(Continued)

Type Field	Description
11	Time exceeded (datagram)
12	Parameter problem
13	(Datagram)
14	Timestamp request
15	Timestamp reply
16	Address mask request
17	Address mask reply
18	

Table 9.2
ICMP message types

Redirection

When a gateway (router) detects that a source node is not using the best route in which to transmit its datagram, it sends a message to the node advising it of the better route.

The code values are shown in Table 9.3.

Code	Description
0	Not used
1	Redirect datagrams for the host node
2	Redirect datagrams for the type of service and net
3	Redirect datagrams for the type of service and host

Table 9.3
Redirection code values

Time exceeded

If a datagram has traversed too many routers, its TTL counter will eventually reach a count of zero. The ICMP time-exceeded message is then sent back to the source node. The time-exceeded message will also be generated if one of the fragments of a fragmented datagram fails to arrive at the destination node Table 9.4.

Code	Description
0	Time to live count exceeded
1	Fragment re-assembly time exceeded

Table 9.4
Time exceeded code values

Parameter problem messages

When there are problems with a particular datagram's contents, a parameter problem message is sent to the original source. The pointer field points to the problem bytes.

Unreachable destination

When a gateway is unable to deliver a datagram, it responds with this message. The datagram is then 'dropped' (deleted) Table 9.5.

Code	Description
0	Network unreachable
1	Host unreachable
2	Protocol unreachable
3	Port unreachable
4	Fragmentation needed and DF set
5	Source route failed
6	Destination network unknown
7	Destination node unknown
8	Source host isolated
9	Communication with destination network prohibited
10	Communication with destination node prohibited
11	Network unreachable for type of service
12	Host unreachable for type of service

Table 9.5
Unreachable code messages

In addition to the reports on errors and exceptional conditions, there is a set of ICMP messages to request information, and to reply to such request.

Echo request and reply

An echo request message is sent to the destination node. This message essentially inquires: 'Are you alive?' A reply indicates that the pathway (i.e. the network(s) in between, the gateways (routers) and the destination node) are all operating correctly.

Timestamp request and replies

This can be used to estimate to synchronize the clock of a host with that of a timeserver.

Sub-net mask request and reply

This can be used by a host to obtain the correct sub-net mask. Where implemented, one or more hosts in the internetwork are designated as subnet mask servers and run a process that replies to sub-net mask request messages.

9.3.3   Routing

Unlike the host-to-host layer protocols (e.g. TCP), which control end-to-end communications, IP is rather 'short-sighted'. Any given IP node (host or router) is only concerned with routing (switching) the datagram to the next node, where the process is repeated. Very few routers have knowledge about the entire internetwork, and often the datagrams are forwarded based on default information without any knowledge of where the destination actually is.

Direct vs indirect delivery

When the source host prepares to send a message to another host, a fundamental decision has to be made, namely: is the destination host also resident on the local network or not? If the NetID portions of the source and destination IP addresses match, the source host will assume that the destination host is resident on the same network, and will attempt to forward it locally using the hardware (Mac) address. This is called direct delivery. If not, the message will be forwarded to the default gateway (a router on the local network), which will forward it using the Mac address. This is called indirect delivery. If the router can deliver it directly i.e. the destination host resides on a network directly connected to the router, it will. If not, it will consult its routing tables and forward it to the next appropriate router. This process will repeat itself until the packet is delivered to its final destination (Figure 9.8).

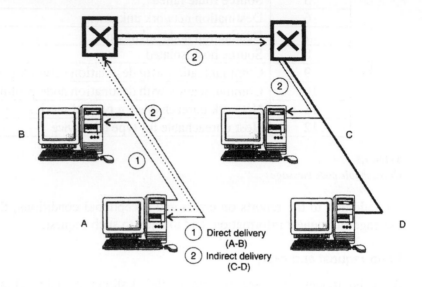

Figure 9.8
Direct vs indirect delivery

Static vs dynamic routing

Each router maintains a table with the following format:
Active routes:

Network Address	Netmask	Gateway Address	Interface	Metric
127.0.0.0	255.0.0.0	127.0.0.1	127.0.0.1	1
207.194.66.0	255.255.255.224	207.194.66.100	207.194.66.100	1
207.194.66.0	255.255.255.255	127.0.0.1	127.0.0.1	1
207.194.66.255	255.255.255.255	207.194.66.100	207.194.66.100	1
224.0.0.0	224.0.0.0	207.194.66.100	207.194.66.100	1
255.255.255.255	255.255.255.255	207.194.66.100	0.0.0.0	1

It basically reads as follows: 'If a packet is destined for network 207.194.66.0, with a netmask of 255.255.255.224, then forward it to the router port 207.194.66.100', etc. It is logical that a given router cannot contain the whereabouts of each and every network in the world in its routing tables, hence it will contain default routes as well. If a packet cannot be specifically routed, it will be forwarded on a default route which should (it is hoped) move it closer to its intended destination.

These routing tables can be maintained in two ways. In most cases, the routing protocols will do this automatically. The routing protocols are implemented in software that runs on the routers, enabling them to communicate on a regular basis and allowing them to share their 'knowledge' about the network with each other. In this way they continuously 'learn' about the topology of the system, and upgrade their routing tables accordingly. This process is called dynamic routing. If, for example, a particular router is removed from the system, the routing tables of all routers containing a reference to that router will change. However, because of the interdependence of the routing tables, a change in any given table will initiate a change in many other routers and it will be a while before the tables stabilize. This process is known as convergence.

Dynamic routing can be further subclassified as distance vector, link-state, or hybrid – depending on the method by which the routers calculate the optimum path.

In distance vector dynamic routing, the 'metric' or yardstick used for calculating the optimum routes is simply based on distance, i.e. which route results in the least number of 'hops' to the destination. Each router constructs a table, which indicates the number of hops to each known network. It then periodically passes copies of its tables to its immediate neighbors. Each recipient of the message then simply adjusts its own tables based on the information received from its neighbor.

The major problem with the distance vector algorithm is that it takes some time to converge to a new understanding of the network. The bandwidth and traffic requirements of this algorithm can also affect the performance of the network, since all routers are periodically exchanging routing tables with each other. The major advantage of the distance vector algorithm is that it is simple to configure and maintain as it only uses the distance to calculate the optimum route.

Link-state routing protocols are also known as shortest path first protocols. This is based on the routers exchanging 'link-state advertisements' to the other routers. Link state advertisement messages contain information about error rates and traffic densities and are triggered by events rather than running periodically as with the distance routing algorithms. Routing decisions are not based on hops alone, but also on the other factors mentioned above.

Hybridized routing protocols use both the methods described above and are more accurate than the conventional distance vector protocols. They converge more rapidly to an understanding of the network than distance vector protocols and avoid the overheads of the link-state updates.

It is also possible for a network administrator to make static entries into routing tables. These entries will not change, even if a router that they point to is not operational.

Autonomous systems

For the purpose of routing a TCP/IP-based internetwork can be divided into several autonomous systems (ASs) or domains. An AS consists of hosts, routers and data links that form several physical networks and are administered by a single authority such as a service provider, university, corporation, or government agency.

Routing decisions that are made within an (AS) are totally under the control of the administering organization. Any routing protocol, using any type of routing algorithm, can be used within an AS since the routing between two hosts in the system are completely isolated from any routing that occurs in other ASs. Only if a host within one AS communicates with a host outside the system, will another AS (or systems) and possibly the Internet backbone be involved.

An AS can be classified under one of three categories:

1. *Stub AS*: This is an AS that has only one connection to the 'outside world' and therefore does not carry any third-party traffic. This is typical of a smaller corporate network.
2. *Multihomed non-transit AS*: This is an AS that has two or more connections to the 'outside world' but is not set up to carry any third party traffic. This is typical of a larger corporate network.
3. *Transit AS*: This is an AS with two or more connections to the outside world, and is set up to carry third party traffic. This is typical of an ISP network.

Interior, exterior and gateway to gateway protocols

There are three categories of TCP/IP routing protocols namely interior gateway protocols, exterior gateway protocols, and gateway to gateway protocols.

Two routers that communicate directly with one another and are both part of the same AS are said to be interior neighbors and are called interior gateways. They communicate with each other using interior gateway protocols (Figure 9.9).

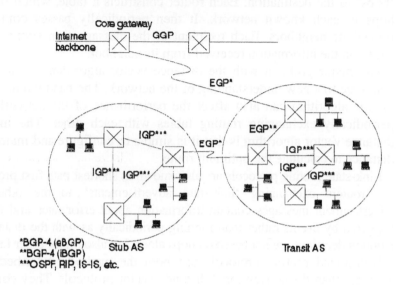

Figure 9.9
Autonomous systems and routing protocols

In a simple AS consisting of only a few physical networks, the routing function provided by IP may be sufficient. In larger ASs, however, sophisticated routers using adaptive routing algorithms may be needed. These routers will communicate with each other using interior gateway protocols such as RIP, HELLO, IS-IS or OSPF.

Routers in different ASs, however, cannot use IGPs for communication for more than one reason. Firstly, IGPs are not optimized for long-distance path determination. Secondly, the owners of ASs (particularly Internet service providers) would find it unacceptable for their routing metrics (which include sensitive information such as error rates and network traffic) to be visible to their competitors. For this reason routers that communicate with each other and are resident in different ASs communicate with each other using exterior gateway protocols.

The routers on the periphery, connected to other ASs, must be capable of handling both the appropriate IGPs and EGPs.

The most common exterior gateway protocol currently used in the TCP/IP environment is border gateway patrol (BGP), the current version being BGP-4.

A third type of routing protocol us used by the core routers (gateways) that connect users to the Internet backbone. They use gateway to gateway protocols (GGP) to communicate with each other.

Interior gateway protocols

The protocols that will be discussed are RIP (routing information protocol) and OSPF (open shortest path first).

RIP

RIP (RFC 1058, 1388) is one of the oldest routing protocols. The original RIP could not handle variable-length sub-net masks, and hence could not support classless interdomain routing (CIDR). This capability has been included with RIPv2.

RIPv2 is a distance vector routing protocol where each router, using a special packet to collect and share information about distances, keeps a routing table of its perspective of the network showing the number of hops required to reach each network. RIP uses as a metric the hop counts.

The RIP routers have fixed update intervals and each router broadcasts its entire routing table to other routers at 30 s intervals (60 s for netware RIP). Each router takes the routing information from its neighbor, adds one hop to the route to account for itself, and then broadcasts its updated table. If an entry has not been updated within 180 s it is assumed suspect and the hop field set to 16 to mark the route as unreachable and it is later removed from the routing table.

One of the major problems with distance vector protocols like RIP is the convergence time, which is the time it takes for the routing information on all routers to settle in response to some change to the network. For a large network the convergence time can be long and there is a greater chance of frames being misrouted.

OSPF

OSPF (RFC 1131, 1247, 1583) is a link state routing or shortest path first protocol designed specifically as an IP routing protocol; hence it cannot transport IPX or AppleTalk protocols. It is encapsulated directly in the IP protocol. OSPF can quickly detect topological changes by flooding link state advertisements to all the other neighbors with reasonably quick convergence.

Each router periodically uses a broadcast mechanism to transmit information to all other routers about its own directly connected routers and the status of the data links to them. Based on the information received from all the other routers each router then constructs its own network routing tree using the shortest path algorithm. All routers continually monitor the status of their links by sending packets to neighboring routers. When the status of a router or link changes, this information is broadcast to the other routers that then update their routing tables. This process is known as flooding and the packets sent are very small representing only the link state changes.

Using cost as the metric OSPF can support a much larger network than RIP, which is limited to 15 routers. A problem area can be in mixed RIP and OSPF environments if routers go from RIP to OSPF and back when hop counts are not incremented correctly.

Exterior gateway protocols

One of the earlier EGPs was in fact called EGP. The current de facto Internet standard for interdomain (AS) routing is Border Gateway Patrol version 4, or simply BGP-4.

BGP-4

BGP-4, as detailed in RFC 1771, performs intelligent route selection based on the shortest AS path. In other words, whereas IGPs such as RIP make decisions on the number of routers to a specific destination, BGP-4 bases its decisions on the number of ASs to a specific destination. It is a so-called path vector protocol, and runs over TCP (port 179).

BGP routers in one AS speak BGP to routers in other ASs, where the 'other' AS might be that of an Internet service provider, or another corporation. Companies with an international presence and a large, global WAN, may also opt to have a separate AS on each continent (running OSPF internally) and run BGP between them in order to create a clean separation.

GGP comes in two 'flavors' namely 'internal' BGP (iBGP) and 'external BGP' (eBGP). IBGP is used within an AS and eBGP between ASs. In order to ascertain which one is used between two adjacent routers, one should look at the AS number for each router. BGP uses a formally registered AS number for entities that will advertize their presence in the Internet. Therefore, if two routers share the same AS number, they are probably using iBGP and if they differ, the routers speak eBGP. BGP routers are referred to as 'BGP speakers', all BGP routers are 'peers', and two adjacent BGP speakers are 'neighbors'.

As mentioned earlier, iBGP is the form of BGP that exchanges BGP updates within an AS. Before information is exchanged with an external AS, iBGP ensures that networks within the AS are reachable. This is done by a combination of 'peering' between BGP routers within the AS and by distributing BGP routing information to IGPs that run within the AS, such as EIGRP, IS-IS, RIP or OSPF. Within the AS, BGP peers do not have to be directly connected as long as there is an IGP running between them. The routing information exchanged consists of a series of AS numbers that describe the full path to the destination network. This information is used by BGP to construct a loop-free map of the network.

In contrast with iBGP, eBGP handles traffic between routers located on different ASs. It can do load balancing in the case of multiple paths between two routers. It also has a synchronization function that, if enabled, will prevent a BGP router from forwarding remote traffic to a transit AS before it has been established that all internal non-BGP routers within that AS are aware of the correct routing information. This is to ensure that packets are not dropped in transit through the AS.

9.3.4   Host-to-host layer: end to end reliability

TCP

Transmission control protocol (TCP) is a connection-oriented protocol and is said to be 'reliable', although this word is used in a data communications context. TCP establishes a session between two machines before data is transmitted. Because a connection is set up beforehand, it is possible to verify that all packets are received on the other end and to arrange retransmission in case of lost packets. Because of all these built-in functions, TCP involves significant additional overhead in terms of processing time and header size.

TCP fragments large chunks of data into smaller segments if necessary, reconstructs the data stream from packets received, issues acknowledgment of data received, provides socket services for multiple connections to ports on remote hosts, performs packet verification and error control, and performs flow control.

The TCP header is structured as in Figure 9.10.

The source and destination ports (16 bits each) identify the host processes at each side of the connection. Examples are post office protocol (POP3) at port 110 and simple mail transfer protocol (SMTP) at port 25. Whereas a destination host is identified by its IP address, the process on that host is identified by its port number. A combination of port number and IP address is called a socket.

The sequence number (32 bits) ensures the sequentiality of the data stream. TCP by implication associates a 32-bit number with every byte it transmits. The sequence number is the number of the first byte in every segment (or 'chunk') of data sent by TCP. If the SYN flag is set, however, it indicates that the sender wants to establish a connection and the number in the sequence number field becomes the initial sequence number or ISN. The receiver acknowledges this, and the sender then labels the first byte of the transmitted data with a sequence number of ISN + 1. The ISN is a pseudo-random number with values between 0 and 2^{32}.

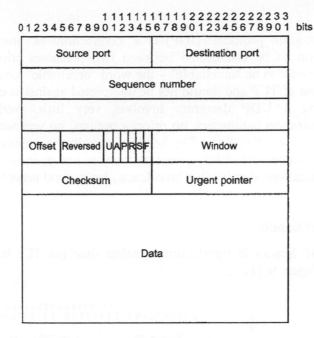

Figure 9.10
TCP header format

The acknowledgment number (32 bits) is used to verify correct receipt of the transmitted data. The receiver checks the incoming data and if the verification is positive, acknowledges it by placing the number of the next byte expected in the acknowledgment number field and setting the ACK flag. The sender, when transmitting, sets a timer and if acknowledgment is not received within a specific time, an error is assumed and the data is retransmitted.

Data offset (4 bits) is the number of 32-bit words in the TCP header. This indicates where the data begins. It is necessary since the header can contain options and thus does not have a fixed length.

Six flags control the connection and data transfer. They are:

1. URG: Urgent pointer field significant
2. ACK: Acknowledgment field significant
3. PSH: Push function
4. RST: Reset the connection
5. SYN: Synchronize sequence numbers
6. FIN: No more data from sender.

The Window field (16 bits) provides flow control. Whenever a host sends an acknowledgment to the other party in the bidirectional communication, it also sends a Window advertisement by placing a number in the Window field. The Window size indicates the number of bytes, starting with the one in the acknowledgment field, that the host is able to accept.

The checksum field (16 bits) is used for error control.

The urgent pointer field (16 bits) is used in conjunction with the URG flag and allows for the insertion of a block of 'urgent' data in the beginning of a particular segment. The pointer points to the first byte of the non-urgent data following the urgent data.

UDP

User datagram protocol (UDP) is a 'connectionless' protocol and does not require a connection to be established between two machines prior to data transmission. It is therefore said to be 'unreliable' – the word 'unreliable' used here as opposed to 'reliable' in the case of TCP and should not be interpreted against its everyday context.

Sending a UDP datagram involves very little overhead in that there are no synchronization parameters, no priority options, no sequence numbers, no timers, and no retransmission of packets. The header is small, the protocol is streamlined functionally. The only major drawback is that delivery is not guaranteed. UDP is therefore used for communications that involve broadcasts, for general network announcements, or for real-time data.

The UDP header

The UDP header is significantly smaller than the TCP header and only contains four fields (Figure 9.11).

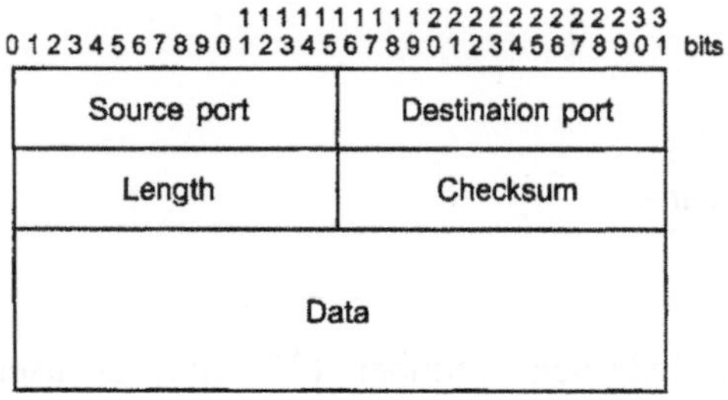

```
                         1 1 1 1 1 1 1 1 1 1 2 2 2 2 2 2 2 2 2 2 3 3
         0 1 2 3 4 5 6 7 8 9 0 1 2 3 4 5 6 7 8 9 0 1 2 3 4 5 6 7 8 9 0 1  bits
```

Source port	Destination port
Length	Checksum
Data	

Figure 9.11
UDP header format

UDP source port (16 bits) is an optional field. When meaningful, it indicates the port of the sending process, and may be assumed to be the port to which a reply should be addressed in the absence of any other information. If not used, a value of zero is inserted.

UDP destination port has the same meaning as for the TCP header, and indicates the process on the destination host to which the data is to be sent.

UDP message length is the length in bytes of the datagram including the header and the data.

The 16-bit UDP checksum is used for validation purposes.

9.3.5   Protocols supporting VoIP

Multicast IP

Normally IP is used for unicasting, i.e. messages are delivered on a one-to-one basis. IP multicasting delivers data on a one-to-many basis simultaneously. To support multicasting, the end stations must support the Internet group management protocol, IGMP, (RFC 1112), which enables multicast routers to determine which end stations on their attached Subnets are members of multicast groups. IPv4 Class D addresses ranging from 224.0.0.0 to 239.255.255.255 are reserved for multicasting.

In order to deliver the multicast datagrams, the routers involved have to use modified routing protocols, such as distance vector multicast routing protocol (DVMRP) and multicast extensions to OSPF (MOSPF).

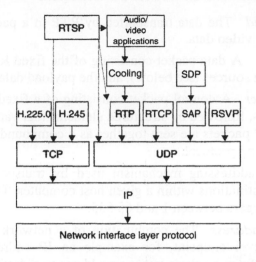

Figure 9.12
VoIP protocols

RTP

As shown earlier, IP provides a connectionless (unreliable) service between two hosts. It does not confirm delivery at all, and cannot deliver packets on a real-time basis. It is up to a protocol at a higher level to determine whether the packets have arrived at their destination at all. There is an added complication in the case of real-time data such as voice and video, which require a deterministic type of delivery.

This problem is addressed by the real-time transport protocol (RTP), an application layer protocol described in RFC1889, which provides end-to-end delivery services such as sequence numbering and stamping for data with real-time characteristics such as interactive audio and video. User applications typically run RTP on top of UDP to make use of its multiplexing and checksum services, but RTP may also be used with other suitable underlying network or transport protocols.

RTP itself does not provide any mechanism to ensure timely delivery or provide other quality-of-service guarantees, but relies on lower-layer services to do so. It does not guarantee delivery or prevent out-of-order delivery, nor does it assume that the underlying network is reliable or that it delivers packets in sequence. The sequence numbers included in RTP allow the receiver to reconstruct the sender's packet sequence, but sequence numbers might also be used to determine the proper location of a packet, for example in video decoding, without necessarily decoding packets in sequence.

RTP consists of two parts, namely the real-time transport protocol (RTP), which carries real-time data, and the RTP control protocol (RTCP), which monitors the quality of service and conveys information about the participants in a session.

The RTP specification is deliberately not complete. In addition, a complete specification of RTP for a particular application will require one or more companion documents such as (a) a profile specification document, which defines a set of payload type codes and their mapping to payload formats (e.g., media encoding), and (b) payload format specification documents, which define how a particular payload, such as an audio or video encoding, is to be carried in RTP.

Definitions

The following RTP definitions are given in RFC 1889. It is necessary to understand them before trying to understand the RTP header.

RTP payload The data transported by RTP in a packet, for example audio samples or compressed video data.

RTP packet A data packet consisting of the fixed RTP header, a possibly empty list of contributing sources (see below), and the payload data.

RTCP packet A control packet consisting of a fixed header part similar to that of RTP data packets, followed by structured elements that vary depending upon the RTCP packet type. RTCP packets are sent together as a compound RTCP packet in a single packet of the underlying protocol.

Port The addressing mechanism used by transport protocols to distinguish among multiple destinations within a given host computer. TCP/IP protocols identify ports using positive integers between 1 and 65 535.

Transport address The combination of a network address and port that identifies a transport level endpoint, for example an IP address and a UDP port. Packets are transmitted from a source transport address to a destination transport address. In TCP/IP jargon this refers to a socket.

RTP session The association among a set of participants communicating with RTP. For each participant, the session is defined by a particular pair of destination transport addresses (one network address plus a port pair for RTP and RTCP). The destination transport address pair may be common for all participants, as in the case of IP multicast, or may be different for each, as in the case of individual unicast network addresses plus a common port pair. In a multimedia session, each medium is carried in a separate RTP session with its own RTCP packets. The multiple RTP sessions are distinguished by different port number pairs and/or different multicast addresses.

Synchronization source (SSRC) The source of a stream of RTP packets, identified by a 32-bit numeric SSRC identifier carried in the RTP header so as not to be dependent upon the network address. All packets from a synchronization source form part of the same timing and sequence number space, so a receiver groups packets by synchronization source for playback. Examples of synchronization sources include the sender of a stream of packets derived from a signal source such as a microphone or a camera, or an RTP mixer (as defined below). A synchronization source may change its data format, e.g., audio encoding, over time. The SSRC identifier is a randomly chosen value meant to be globally unique within a particular RTP session.

Contributing source (CSRC) A source of a stream of RTP packets that has contributed to the combined stream produced by an RTP mixer. The mixer inserts a list of the SSRC identifiers of the sources that contributed to the generation of a particular packet into the RTP header of that packet. This list is called the CSRC list. An example application is audio conferencing where a mixer indicates all the talkers whose speech was combined to produce the outgoing packet, allowing the receiver to indicate the current talker, even though all the audio packets contain the same SSRC identifier (that of the mixer).

End system An application that generates the content to be sent in RTP packets and/or consumes the content of received RTP packets. An end system can act as one or more synchronization sources in a particular RTP session, but typically only one.

Mixer An intermediate system that receives RTP packets from one or more sources, possibly changes the data format, combines the packets in some manner and then forwards a new RTP packet. Since the timing among multiple input sources will not generally be synchronized, the mixer will make timing adjustments among the streams and generate its own timing for the combined stream. Thus, all data packets originating from a mixer will be identified as having the mixer as their synchronization source.

Translator An intermediate system that forwards RTP packets with their synchronization source identifier intact. Examples of translators include devices that convert encodings without mixing, replicators from multicast to unicast, and application level filters in firewalls.

Monitor An application that receives RTCP packets sent by participants in an RTP session, in particular the reception reports, and estimates the current quality of service for distribution monitoring, fault diagnosis and long-term statistics. The monitor function is likely to be built into the application(s) participating in the session, but may also be a separate application that does not otherwise participate and does not send or receive the RTP data packets. These are called third party monitors.

Non-RTP means Protocols and mechanisms that may be needed in addition to RTP to provide a usable service. In particular, for multimedia conferences, a conference control application may distribute multicast addresses and keys for encryption, negotiate the encryption algorithm to be used, and define dynamic mappings between RTP payload type values and the payload formats they represent for formats that do not have a pre-defined payload type value. For simple applications, electronic mail or a conference database may also be used.

The RTP header format

The first section (12 bytes fixed) is present in all RTP headers. The second section is optional (and variable length) and consists of CSRC identifiers inserted by a mixer. The third section is also optional and variable in length, and is used for experimental purposes (Figure 9.13).

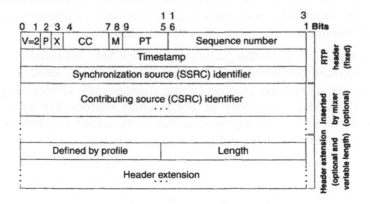

Figure 9.13
RTP header

Version (V) 2 bits. This field identifies the version of RTP. The current version is 2.

Padding (P) 1 bit. If the padding bit is set, the packet contains one or more additional padding octets at the end which are not part of the payload. The last octet of the padding contains a count of how many padding octets should be ignored. Padding may be needed by some encryption algorithms with fixed block sizes or for carrying several RTP packets in a lower layer protocol data unit.

Extension (X) 1 bit. If the extension bit is set, the fixed header is followed by exactly one header extension, with a format defined in RFC 1889 Section 5.3.1.

CSRC count (CC) 4 bits. The CSRC count contains the number of CSRC identifiers that follow the fixed header.

Marker (M)   1 bit. The interpretation of the marker is defined by a profile. It is intended to allow significant events such as frame boundaries to be marked in the packet stream.

Payload type (PT)   7 bits. This field identifies the format of the RTP payload and determines its interpretation by the application. A profile specifies a default static mapping of payload type codes to payload formats.

Sequence number   16 bits. The sequence number increments by one for each RTP data packet sent, and may be used by the receiver to detect packet loss and to restore packet sequence. The initial value of the sequence number is random (unpredictable) to make known plaintext attacks on encryption more difficult, even if the source itself does not encrypt, because the packets may flow through a translator that does.

Timestamp   32 bits. The timestamp reflects the sampling instant of the first octet in the RTP data packet. The sampling instant must be derived from a clock that increments monotonically and linearly in time to allow synchronization and jitter calculations. The resolution of the clock must be sufficient for the desired synchronization accuracy and for measuring packet arrival jitter (one tick per video frame is typically not sufficient). The clock frequency is dependent on the format of data carried as payload and is specified statically in the profile or payload format specification that defines the format, or may be specified dynamically for payload formats defined through non-RTP means. The initial value of the timestamp is random, as for the sequence number. Several consecutive RTP packets may have equal timestamps if they are (logically) generated at once, e.g., belong to the same video frame. Consecutive RTP packets may contain timestamps that are not monotonic if the data is not transmitted in the order it was sampled, as in the case of MPEG-interpolated video frames.

SSRC   32 bits. The SSRC field identifies the synchronization source. This identifier is chosen randomly, with the intent that no two synchronization sources within the same RTP session will have the same SSRC identifier.

CSRC list   0–15 items, 32 bits each. The CSRC list identifies the contributing sources for the payload contained in this packet. The number of identifiers is given by the CC field. If there are more than 15 contributing sources, only 15 may be identified. CSRC identifiers are the SSRC identifiers of contributing sources. For example, for audio packets the SSRC identifiers of all sources that were mixed together to create a packet are listed, allowing correct talker indication at the receiver.

RTP header extension   An extension mechanism is provided to allow individual implementations to experiment with new payload-format-independent functions that require additional information to be carried in the RTP data packet header.

RTCP

The RTCP is based on the periodic transmission of control packets to all participants in the session, using the same distribution mechanism as the data packets. The underlying protocol must provide multiplexing of the data and control packets, for example using separate port numbers with UDP. RTCP performs the following functions.

The primary function is to provide feedback on the quality of the data distribution. This is an integral part of RTP's role as a transport protocol and is related to the flow and congestion control functions of other transport protocols. The feedback may be directly useful for control of adaptive encoding, but experiments with IP multicasting have shown that it is also critical to get feedback from the receivers to diagnose faults in the distribution. Sending reception feedback reports to all participants allows one who is observing problems to evaluate whether those problems are local or global. With a distribution mechanism like IP multicast, it is also possible for an entity such as a network service provider who is not otherwise involved in the session to receive the feedback information and act as a third-party monitor to diagnose network problems. This feedback function is performed by the RTCP sender and receiver reports, described in RFC 1889 Section 6.3.

RTCP carries a persistent transport-level identifier for an RTP source called the canonical name or CNAME. Since the SSRC identifier may change if a conflict is discovered or a program is restarted, receivers require the CNAME to keep track of each participant. Receivers also require the CNAME to associate multiple data streams from a given participant in a set of related RTP sessions, for example to synchronize audio and video.

The first two functions require that all participants send RTCP packets, therefore the rate must be controlled in order for RTP to scale up to a large number of participants. By having each participant send its control packets to all the others, each can independently observe the number of participants. This number is used to calculate the rate at which the packets are sent.

A fourth, optional function is to convey minimal session control information, for example participant identification to be displayed in the user interface. This is most likely to be useful in 'loosely controlled' sessions where participants enter and leave without membership control or parameter negotiation. RTCP serves as a convenient channel to reach all the participants, but it is not necessarily expected to support all the control communication requirements of an application.

RTCP packet format

This specification defines several RTCP packet types to carry a variety of control information, namely:

> SR: A sender report, for transmission and reception statistics from participants that are active senders
>
> RR: A receiver report, for reception statistics from participants that are not active senders
>
> SDES: Source description items, including CNAME
>
> BYE: Indicates end of participation
>
> APP: Application specific functions.

Each RTCP packet begins with a fixed part similar to that of RTP data packets, followed by structured elements that may be of variable length according to the packet type but always end on a 32-bit boundary. The alignment requirement and a length field in the fixed part are included to make RTCP packets 'stackable'. Multiple RTCP packets may be concatenated without any intervening separators to form a compound RTCP packet that is sent in a single packet of the lower layer protocol, for example UDP. There is no explicit count of individual RTCP packets in the compound packet since the lower layer protocols are expected to provide an overall length to determine the end of the compound packet.

For further details on RTCP refer to RFC 1889 and RFC 1890.

SDP

The session description protocol, SDP (RFC 2327) is a session description protocol for multimedia sessions. A multimedia session, for these purposes, is defined as a set of media streams that exist for some duration of time. Media streams can be many-to-many and the times during which the session is active need not be continuous.

SDP is used for purposes such as session announcement or session invitation. On the Internet multicast backbone (Mbone), SDP is used to advertize multimedia conferences and communicate the conference addresses and conference tool-specific information necessary for participation. The Mbone is the part of the Internet that supports IP multicast, and thus permits efficient many-to-many communication. It is used extensively for multimedia conferencing.

SDP is purely a format for session description. It does not incorporate a transport protocol, and has to use different transport protocols as appropriate including the session announcement protocol (SAP), session initiation protocol (SIP), real-time streaming protocol (RTSP), electronic mail using the MIME extensions, and the hypertext transport protocol (HTTP).

A common approach is for a client to announce a conference session by periodically multicasting an announcement packet to a well-known multicast address and port using the Session announcement protocol (SAP). An SAP packet consists of an SAP header with a text payload, which is an SDP session description no greater than 1 kb in length. If announced by SAP, only one session announcement is permitted in a single packet.

Alternative means of conveying session descriptions include e-mail and the World Wide Web (WWW). For both e-mail and WWW distribution, the use of the MIME content type 'application/sdp' is used. This enables the automatic launching of applications for participation in the session from the WWW client or mail reader.

SDP serves two primary purposes. It is a means to communicate the existence of a session, as well as a means to convey sufficient information to enable joining and participating in the session. In this regard SDP includes information such as the session name and purpose, the time(s) the session is active, the media comprising the session, information to receive those media (addresses, ports, formats, etc.), information about the bandwidth to be used by the conference and contact information for the person responsible for the session.

Detailed information regarding the media include the type of media (video, audio, etc.), the transport protocol (RTP/UDP/IP, H.320, etc.), as well as the format of the media (H.261 video, MPEG video, etc.).

For an IP multicast session, the multicast address as well as the port number for media are supplied. This IP address and port number are the destination address and destination port of the multicast stream, whether being sent, received, or both. For an IP unicast session, the remote IP address for media as well as the port number for the contact address is supplied.

Sessions may either be bounded or unbounded in time. Whether or not they are bounded, they may be only active at specific times. With regard to timing information, SDP can convey an arbitrary list of start and stop times bounding the session. It can also include repeat times for each bound, such as 'every Wednesday at 10 am for one hour'. This timing information is globally consistent, irrespective of local time zone or daylight saving time.

RTSP

The real-time streaming protocol (RTSP) is an application layer protocol that establishes and controls either a single or several time-synchronized streams of continuous media such as audio and video. It normally does not deliver the continuous streams itself, although interleaving of the continuous media stream with the control stream is possible. In other words, RTSP acts as a 'network remote control' for multimedia servers.

There is no notion of an RTSP connection. Instead, a server maintains a session labeled by an identifier. An RTSP session is not tied to a transport-level connection such as a

TCP connection. During an RTSP session, an RTSP client may open and close many reliable transport connections to the server to issue RTSP requests. Alternatively, it may use a connectionless transport protocol such as UDP.

The streams controlled by RTSP may use RTP, but the operation of RTSP does not depend on the transport mechanism used to carry continuous media. The protocol is intentionally similar in syntax and operation to HTTP version 1.1 so that extension mechanisms to HTTP can in most cases also be added to RTSP. However, RTSP differs fundamentally from HTTP/1.1 in that data delivery takes place 'out-of-band' in a different protocol. HTTP is an asymmetric protocol where the client issues requests and the server responds. In RTSP, both the media client and media server can issue requests. RTSP requests are also not stateless; they may set parameters and continue to control a media stream long after the request has been acknowledged.

The protocol supports the following operations:

- *Retrieval of media from media server*: The client can request a presentation description via HTTP or some other method. If the presentation is being multicast, the presentation description contains the multicast addresses and ports to be used for the continuous media. If the presentation is to be sent only to the client via unicast, the client provides the destination for security reasons.
- *Invitation of a media server to a conference*: A media server can be 'invited' to join an existing conference, either to playback media into the presentation or to record all or a subset of the media in a presentation. This mode is useful for distributed teaching applications. Several parties in the conference may take turns 'pushing the remote control buttons'.
- *Addition of media to an existing presentation*: Particularly for live presentations, it is useful if the server can tell the client about additional media becoming available.

RSVP

The need for a quality of service (QoS) on the Internet arose because of the increasing number of time-sensitive applications involving voice and video. RSVP (RFC 2205) is a resource reservation set-up protocol designed for that purpose. It is used by a host to request specific qualities of service from the network for particular application data streams or flows. It is also used by routers to deliver quality-of-service (QoS) requests to all nodes along the path(s) of the flows and to establish and maintain the requested service. RSVP requests will generally result in resources being reserved in each node along the data path.

RSVP requests resources for 'simplex' flows, i.e. it requests resources in only one direction. Therefore, RSVP treats a sender as logically distinct from a receiver, although the same application process may act as both a sender and a receiver at the same time. RSVP operates on top of IPv4 or IPv6, occupying the place of a transport protocol in the protocol stack. However, RSVP does not transport application data but is rather an Internet control protocol, like ICMP, IGMP, or routing protocols. Like the implementations of routing and management protocols, an implementation of RSVP will typically execute in the background, not in the data path.

RSVP is not itself a routing protocol. Instead, it is designed to operate with current and future unicast and multicast routing protocols. An RSVP process consults the local routing database(s) to obtain routes. In the multicast case, for example, a host sends IGMP messages to join a multicast group and then sends RSVP messages to reserve resources along the delivery path(s) of that group. Whereas routing protocols determine where packets get forwarded, RSVP is only concerned with the QoS of those packets that are forwarded in accordance with routing decisions.

Quality of service is implemented for a particular data flow by mechanisms collectively called 'traffic control'. These mechanisms include (1) a packet classifier, (2) admission control and (3) a 'packet scheduler' or some other datalink-layer-dependent mechanism to determine when particular packets are forwarded. The 'packet classifier' determines the QoS class (and perhaps the route) for each packet. For each outgoing interface, the 'packet scheduler' or other datalink-layer-dependent mechanism achieves the promised QoS. Traffic control implements QoS service models defined by the Integrated Services Working Group.

In summary, RSVP has the following attributes:

- It makes resource reservations for both unicast and many-to-many multicast applications, adapting dynamically to changing group membership as well as to changing routes.
- It is simplex, i.e., it makes reservations for unidirectional data flows.
- It is receiver-oriented, i.e., the receiver of a data flow initiates and maintains the resource reservation used for that flow.
- It maintains 'soft' state in routers and hosts, providing support for dynamic membership changes and automatic adaptation to routing changes.
- It is not a routing protocol but depends upon present and future routing protocols.
- It transports and maintains traffic control and policy control parameters that are opaque to RSVP.
- It provides several reservation models or 'styles' to fit a variety of applications.
- It provides transparent operation through routers that do not support it.
- It supports both IPv4 and IPv6.

The RSVP message consists of three sections namely a common header (8 bytes), an object header (4 bits) and object contents (variable length).

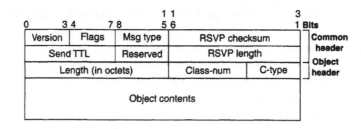

Figure 9.14
RSVP header

The version field (4 bits) contain the current version which is 2.
The flags field (4 bits) is reserved for future use.
The message type (1 byte) indicates 1 of 7 currently defined messages. They are:

- *Type 1*: *Path messages* Each sender host periodically sends a path message for each data flow it originates. A path message travels from a sender to receiver(s) along the same path(s) used by the data packets.
- *Type 2*: *Resv messages* Resv (reservation) messages carry reservation requests hop-by-hop from receivers to senders, along the reverse paths of data flows for the session.

- *Type 3*: *PathErr messages* PathErr (path error) messages report errors in processing path messages. They are travel upstream towards senders and are routed hop-by-hop using the path state. PathErr messages do not modify the state of any node through which they pass; they are only reported to the sender application.
- *Type 4*: *ResvErr messages* ResvErr (reservation error) messages report errors in processing resv messages, or they may report the spontaneous disruption of a reservation, e.g., by administrative pre-emption. ResvErr messages travel downstream towards the appropriate receivers, routed hop-by-hop using the reservation state.
- *Type 5/6*: *Teardown messages* RSVP 'teardown' messages remove path or reservation states immediately. Although it is not necessary to explicitly tear down an old reservation, it is recommended that all end hosts send a teardown request as soon as an application finishes. There are two types of RSVP teardown message, namely PathTear (5) and ResvTear (6). A PathTear message travels towards all receivers downstream from its point of initiation and deletes path state, as well as all dependent reservation state, along the way. A ResvTear message deletes reservation state and travels towards all senders upstream from its point of initiation.
- *Type 7*: *ResvConf messages* ResvConf (reservation conformation) messages are sent to acknowledge reservation requests and are forwarded to the receiver hop-by-hop, to accommodate the hop-by-hop integrity check mechanism.

The RSVP checksum field (2 bytes) provides error control.

The send_TTL field (1byte) is the IP time-to-live value with which the message was sent.

The RSVP length field (2 octets) is the total length of the RSVP message in bytes.

Each object consists of one or more 32 bit long words with a 1 byte object header. The object contents are fairly complex and beyond the scope of this document and readers are referred to RFC 2205 for details.

9.4 Summary

Figure 9.15 summarizes this section. It shows a 20 ms voice sample from a coder, preceded by an RTP, UDP, IP and – in this case – an Ethernet header. Note that the drawing is not upside down, but has been drawn according to the actual sequence in which the bytes are transmitted. The operation of the coders and decoders (codecs) are discussed in the following paragraphs.

9.5 Hardware

9.5.1 Introduction

The voice over IP Forum (part of the International Multimedia Teleconferencing Consortium) has developed an Implementation Agreement that describes several connectivity configurations for converged networks. These configurations are summarized in Figure 9.16. It allows for:

- PC to PC connection, as used in Microsoft NetMeeting
- Phone to phone connection over IP as offered by several phone companies via their pre-paid calling card services
- PC to phone connection as used by Net2Phone.

All components shown, with the exception of the domain name (DNS) server, adhere to the H.323 standard.

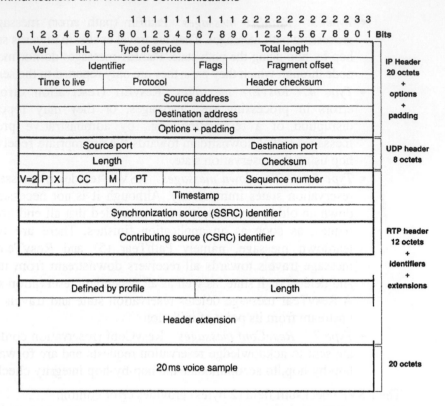

Figure 9.15
VoIP summary

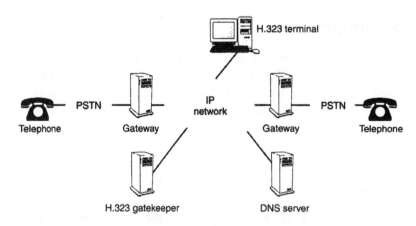

Figure 9.16
Elements used for PC to telephone interconnection

9.5.2   H.323 overview

The ITU-T H.323 (packet-based multimedia communication systems) standard deals with
the components, protocols and procedures required for the transmission of real-time audio,
video, and data communications over packet-based networks including IP or IPX (Internet
packet exchange)-based LANs, MANs, WANs and VPNs. H.323 can be used for audio

only (IP telephony), audio/video (videotelephony), audio/data and audio/video/data. H.323 can also be applied to multipoint-multimedia communications.

The H.323 standard deals with visual telephone systems and equipment for LANs that provide a non-guaranteed quality of service (QoS) and it is part of the H.32x family of ITU-T standards. Other standards in the family include:

- H.320 narrowband visual telephone systems and terminal equipment (used with ISDN)
- H.321 adaptation of H.320 terminals to broadband ISDN (B-ISDN)
- H.322 visual telephone systems and equipment for LANs that provide a guaranteed quality of service
- H.324 terminal for low bit rate multimedia communications (used with PSTN and wireless applications).

In addition to the network implementation standards mentioned above, there are other standards that fall under the umbrella of the H.323 recommendation. These include:

- H.225.0 terminal to gatekeeper functions
- H.245 terminal control functions used to negotiate channel capabilities and usage
- Q.931 call signaling functions to establish and terminate a call
- T.120 data conferencing, including whiteboarding and still image functions.

9.5.3   H.323 building blocks

The H.323 standard specifies four kinds of components, which, when networked together, provide the point-to-point and point-to-multipoint multimedia-communication services. They are:

- Terminals
- Gateways
- Gatekeepers
- Multipoint control units (MCUs).

Terminals

Terminals are used for real-time bidirectional multimedia communications. An H.323 terminal can either be a personal computer (PC) or a stand-alone device, running an H.323 stack and multimedia applications. By default a terminal supports audio communications and it can optionally support video or data communications. H.323 terminals may be used in multipoint conferences.

The primary goal of H.323 is to internetwork multimedia terminals and for this reason H.323 terminals are compatible with H.324 terminals on switched circuit networks (PSTN) and wireless networks, H.310 terminals on B-ISDN, H.320 terminals on ISDN, H.321 terminals on B-ISDN and H.322 terminals on guaranteed QoS LANs.

Gateways

One of the primary goals in the development of the H.323 standard was interoperability between an H.323 network and any other non-H.323 multimedia services network such as a PSTN network. This is achieved through the use of a gateway, which performs any network or signaling translation required between the two networks. This connectivity of dissimilar networks is achieved by translating protocols for call set-up and release, converting media formats between different networks, and transferring information between the networks connected by the gateway.

Gatekeepers

A gatekeeper can be considered the 'brain' of the H.323 network and is the focal point for all calls within the H.323 network. Although not mandatory, gatekeepers provide important services such as addressing, authorization and authentication of terminals and gateways, bandwidth management, accounting, billing, charging and call-routing services.

Multipoint control units

Multipoint control units (MCUs) provide support for conferences of three or more H.323 terminals. All terminals participating in the conference establish a connection with the MCU. The MCU manages conference resources, negotiates between terminals for the purpose of determining the audio or video CODEC to use and may handle the media stream. The gatekeepers, gateways and MCUs are logically separate components of the H.323 standard but can be implemented as a single physical device.

Zones

An H.323 zone is not a building block, but rather a collection of all terminals, gateways, and MCUs managed by a single gatekeeper. A zone includes at least one terminal and may include gateways or MCUs, but only one gatekeeper. A zone is independent of network topology and may be comprised of multiple network segments that are connected using routers, switches or bridges (Figure 9.17).

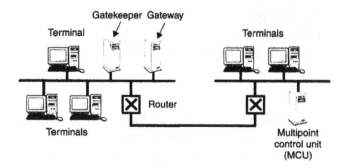

Figure 9.17
H.323 zone

9.5.4   H.323 protocols

The protocols specified by H.323 are listed below. H.323 is, however, independent of the packet network and the transport protocols over which it runs (i.e. the bottom four layers of the OSI model) and does not specify them.

- Audio Codecs
- Video Codecs
- H.225 registration, admission and status (RAS)

- H.225 call signaling
- H.245 control signaling
- Real-time transfer protocol (RTP)
- Real-time control protocol (RTCP).

Here follows brief descriptions of these protocols.

Audio (voice) Codecs

In order to transport the analog voice signal, it has to be digitized. The traditional approach is to sample it at 8000 samples per second, and then to quantize each sample into an 8-bit digital value, with a resolution of $1/(2^8) = 1/256$. Thus the basic bit rate becomes 8000 samples per second $\times$ 8 bits per sample = 64 000 bps, known as the DS0 rate. Through an additional process called encoding, the 64 000 bps bandwidth requirement can be reduced considerably, by employing techniques such as data compression, silence suppression and voice activity detection. On the receiving side the process has to be reversed.

An audio Codec (coder/decoder) encodes the audio signal from the microphone for transmission on the transmitting H.323 terminal and decodes the received audio code that is sent to the speaker on the receiving H.323 terminal. Because audio is the minimum service provided by the H.323 standard, all H.323 terminals must have at least one audio Codec support, as specified in the ITU-T G.711 (1972) recommendation. This algorithm operates at 64 kbps, using PCM to produce a frame that contains 125 µs of speech. No compression is used (Figure 9.18).

Additional Codecs that may be supported include:

- *G.722 (1988):* Operates at 64, 56 or 48 kbps and is known as the wideband coder.
- *G.723.1 (1995):* Operates at 5.3 and 6.3 kbps. Algebraic code excited linear prediction (ACELP) is used for the lower rate and Multipulse maximum likelihood quantization (MP-MLQ) is used for the higher rate.
- *G.726 (1990):* Operates at 16, 24, 32 and 40 kbps, and uses adaptive differential pulse code modulation (ADPCM).
- *G.728 (1995):* Operates at 16 kbps and uses conjugate structure algebraic code excited linear prediction (CS-ACELP).
- *G.729 (1996):* Operates at 8 kbps and uses a less complex version of CS-ACELP.

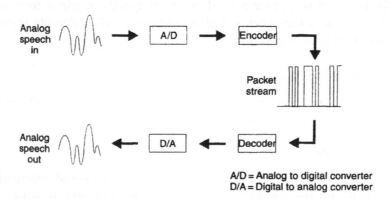

A/D = Analog to digital converter
D/A = Digital to analog converter

Figure 9.18
Packet telephony

Video Codec

A video codec encodes video from a camera for transmission on the transmitting H.323 terminal and decodes the received video code that is sent to the video display on the receiving H.323 terminal. H.323 specifies support of video and hence video Codecs as optional. However, H.323 terminal providing video communications must support video encoding and decoding as specified in the ITU-T H.261 recommendation. H.261 operates at multiples of $64 \times p$ kbps, where p varies from 1 to 30, resulting in bit rates from 40 kbps to 2 Mbps.

An alternative is H.263, based on H.261, which introduces additional compression. It contains negotiable options and can operate on a number of different video formats.

H.225 registration, admission and status

Registration, admission and status (RAS) is the protocol used between H.323 endpoints (terminals/gateways) and gatekeepers for the following:

- Gatekeeper discovery (GRQ)
- Endpoint registration
- Endpoint location
- Admission control.

An RAS signaling channel, opened between an endpoint and a gatekeeper prior to the establishment of any other channels, is used to exchange RAS messages. This channel is unreliable, hence RAS message exchange may be associated with timeouts and retry counts.

Gatekeeper discovery

The gatekeeper discovery process is used by the H.323 endpoints to determine the gatekeeper with which the endpoint must register and can be done statically or dynamically. In static discovery, the endpoint knows the transport address of its gatekeeper a priori. In the dynamic method of gatekeeper discovery, the endpoint multicasts a GRQ message on the gatekeeper's discovery multicast address: 'Who is my gatekeeper?' One or more gatekeepers may respond with a GCF message: 'I can be your gatekeeper.'

Endpoint registration

Registration is a process used by the endpoints to join a zone and inform the gatekeeper of the zone's transport and alias addresses. All endpoints register with a gatekeeper as part of their configuration.

Endpoint location

Endpoint location is a process by which the transport address of an endpoint is determined and given its alias name or E.164 address.

Other control

The RAS channel is used for other kinds of control mechanisms, such as admission control, to restrict the entry of an endpoint into a zone, bandwidth control, and disengagement control, where an endpoint is disassociated from a gatekeeper and its zone.

H.225 call signaling

The H.225 call signaling is used to establish a connection between two H.323 endpoints (terminals/gateways), over which the real-time data can be transported. This is achieved by exchanging H.225 protocol messages on the call-signaling channel. The call-signaling channel is opened between two H.323 endpoints or between an endpoint and the gatekeeper.

Call signaling involves the exchange of H.225 protocol messages over a reliable call-signaling channel, hence H.225 protocol messages are carried over TCP in an IP-based H.323 network. H.225 messages are exchanged between the endpoints if there is no gatekeeper in the H.323 network. When a gatekeeper exists in the network, the H.225 messages are exchanged either directly between the endpoints or between the endpoints after being routed through the gatekeeper. The first case is direct call signaling. The second case is called gatekeeper-routed call signaling. The method chosen is decided by the gatekeeper during RAS-admission message exchange.

Gatekeeper-routed call signaling

The admission messages are exchanged between endpoints and the gatekeeper on RAS channels. The gatekeeper receives the call-signaling messages on the call-signaling channel from one endpoint and routes them to the other endpoint on the call-signaling channel of the other endpoint.

Direct call signaling

During the admission confirmation, the gatekeeper indicates that the endpoints can exchange call-signaling messages directly. The endpoints exchange the call signaling on the call-signaling channel.

H.245 control signaling

H.245 control signaling is used to exchange end-to-end control messages governing the operation of the H.323 endpoints. These control messages carry information related to the following:

- Capabilities exchange
- Opening and closing of logical channels used to carry media streams
- Flow control messages
- General commands and indications.

The H.245 control messages are carried over H.245 control channels. The H.245 control channel is the logical channel 0 and is permanently open, unlike the media channels. The messages carried include messages to exchange capabilities of terminals and to open and close logical channels.

Capabilities exchange

Capabilities exchange is a process using the communicating terminals' exchange messages to provide their transmit and receive capabilities to the peer endpoint. Transmit capabilities describe the terminal's ability to transmit media streams. Receive capabilities describe a terminal's ability to receive and process incoming media streams.

Logical channel signaling

A logical channel carries information from one endpoint to another endpoint (in the case of a point-to-point conference) or multiple endpoints (in the case of a point-to-multipoint conference). H.245 provides messages to open or close a logical channel; a logical channel is unidirectional.

RTP

RTP provides end-to-end delivery services of real-time audio and video using UDP. RTP provides payload-type identification, sequence numbering, time stamping and delivery monitoring. UDP provides multiplexing and checksum services. RTP is discussed in detail in Section 9.3.5.2.

RTCP

RTCP is the counterpart of RTP that provides control services. The primary function of RTCP is to provide feedback on the quality of the data distribution. Other RTCP functions include carrying a transport-level identifier for an RTP source, called a canonical name, which is used by receivers to synchronize audio and video. RTCP is discussed in more detail in Section 9.3.5.3.

9.5.5   Terminal implementation

Terminals implement the following protocols:

- H.245 for exchanging terminal capabilities and creation of media channels
- H.225 for call signaling and call set-up
- RAS for registration and other admission control with a gatekeeper
- RTP/RTCP for sequencing audio and video packets.

H.323 terminals must also support the G.711 audio codec. Optional components in an H.323 terminal are video codecs and T.120 data conferencing protocols (Figure 9.19).

9.5.6   Gateway implementation

A gateway provides translation of protocols for call set-up and release, conversion of media formats between different networks, and the transfer of information between H.323 and non-H.323 networks (Figure 9.20). An application of the H.323 gateway is in IP telephony, where the H.323 gateway connects an IP network and switched circuit networks, e.g. PSTN and ISDN networks.

On the H.323 side, a gateway runs H.245 control signaling for exchanging capabilities, H.225 call signaling for call set-up and release, and H.225 registration, admissions, and status (RAS) for registration with the gatekeeper. On the SCN side, a gateway runs SCN-specific protocols (e.g. ISDN and SS7). Terminals communicate with gateways using the H.245 control-signaling protocol and H.225 call-signaling protocol. The gateway translates these protocols in a transparent fashion to the respective counterparts on the non-H.323 network and vice versa. The gateway also performs call set-up and clearing on both the H.323 network side and the non-H.323 network side. Translation between audio, video and data formats may also be performed by the gateway. Audio and video translation may not be required if both terminal types find a common communications mode. For example, in the case of a gateway to H.320 terminal on ISDN, both terminal types require G.711 audio and H.261 video, so a common mode always exists. The gateway has the characteristics of both an H.323 terminal on the H.323 network and the other terminal on the non-H.323 network it connects. Gatekeepers are aware of which endpoints are gateways because this is indicated when the terminals and gateways register with the gatekeeper. A gateway may be able to support several simultaneous calls between the H.323 and non-H.323 networks. A gateway is a logical component of H.323 and can be implemented as part of a gatekeeper or an MCU.

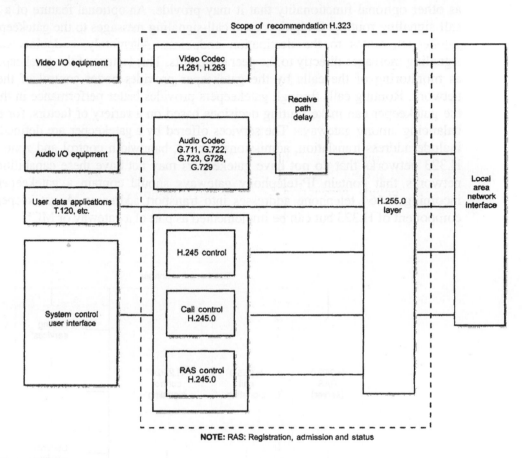

Figure 9.19
Terminal protocol stack

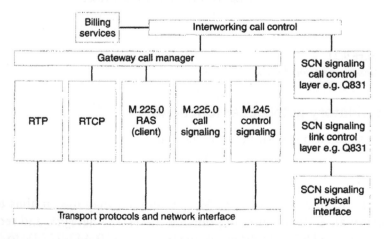

Figure 9.20
Gateway protocol stack

9.5.7   Gatekeeper implementation

Gatekeepers provide call-control services for H.323 endpoints, such as address translation and bandwidth management as defined within RAS and are optional (Figure 9.21). If they are present in a network, however, terminals and gateways must use their services. The H.323 standards define both mandatory services that the gatekeeper must provide, as well as other optional functionality that it may provide. An optional feature of a gatekeeper is call-signaling routing. Endpoints send call-signaling messages to the gatekeeper, which the gatekeeper routes to the destination endpoints. Alternately, endpoints can send call-signaling messages directly to the peer endpoints. This feature of the gatekeeper is valuable, as monitoring of the calls by the gatekeeper provides better control of the calls in the network. Routing calls through gatekeepers provides better performance in the network, as the gatekeeper can make routing decisions based on a variety of factors, for example, load balancing among gateways. The services offered by a gatekeeper are defined by RAS and include address translation, admissions control, bandwidth control and zone management. H.323 networks that do not have gatekeepers may not have these capabilities, but H.323 networks that contain IP-telephony gateways should contain a gatekeeper to translate incoming E.164 telephone addresses into transport addresses. A gatekeeper is a logical component of H.323 but can be implemented as part of a gateway or MCU.

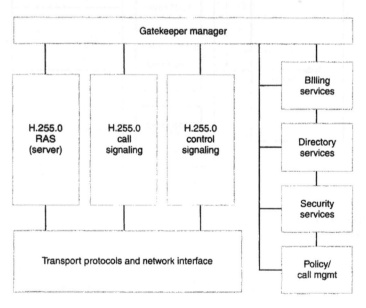

Figure 9.21
Gatekeeper protocol stack

Mandatory gatekeeper functions

These functions *must* be implemented.

Address translation

Calls originating within an H.323 network may use an alias to address the destination terminal. Calls originating outside the H.323 network and received by a gateway may use an E.164 telephone number (e.g. 310-442-9222) to address the destination terminal. The gatekeeper translates this number or the alias into the network address (e.g. 204.252.32.116 for an IP-based network) for the destination terminal. The destination endpoint can be reached using the network address on the H.323 network.

Admissions control

The gatekeeper can control the admission of the endpoints into the H.323 network. It uses RAS messages as specified in H.225.0 such as admission request (ARQ) confirm (ACF) and reject (ARJ) to achieve this. Admissions control may be a null function that admits all endpoints to the H.323 network.

Bandwidth control

The gatekeeper provides support for bandwidth control by using the RAS (H.225.0) messages bandwidth request (BRQ), confirm (BCF) and reject (BRJ). For instance, if a network manager has specified a threshold for the number of simultaneous connections on the H.323 network, the gatekeeper can refuse to make any more connections once the threshold is reached. The result is to limit the total allocated bandwidth to some fraction of the total available, leaving the remaining bandwidth for data applications. Bandwidth control may also be a null function that accepts all requests for bandwidth changes.

Zone management

The gatekeeper provides the above functions – address translation, admissions control and bandwidth control – for terminals, gateways and MCUs located within its zone of control.

Optional gatekeeper functions

Call-control signaling

The gatekeeper can route call-signaling messages between H.323 endpoints. In a point-to-point conference, the gatekeeper may process H.225 call-signaling messages. Alternatively, the gatekeeper may allow the endpoints to send H.225 call-signaling messages directly to each other.

Call authorization

When an endpoint sends call-signaling messages to the gatekeeper, the gatekeeper may accept or reject the call, according to the H.225 specification. The reasons for rejection may include access- or time-based restrictions, to and from particular terminals or gateways.

Call management

The gatekeeper may maintain information about all active H.323 calls so that it can control its zone by providing the maintained information to the bandwidth-management function or by re-routing the calls to different endpoints to achieve load balancing.

9.6 Implementation considerations: quality of service

9.6.1 Definition

Quality of service (QoS) is defined in ITU-T E.800 as 'The collective effect of service performance, which determines the degree of satisfaction of a user of the service.' Unfortunately this is a very subjective evaluation, but there are several techniques for evaluating QoS such as the conversation opinion test outlined in ITU-T P.800. This test uses volunteers to rate the quality of a telephone connection from 1 to 5 where 1 is bad, 2 is poor, 3 is fair, 4 is good and 5 is excellent.

9.6.2   Factors influencing QoS

In a VoIP conversation between a telephone and a PC situated at home, the voice signal has to pass through six distinct phases namely the telephone client, PSTN, gateway, IP network, dial-up link and PC client (Figure 9.22).

Each leg of the journey unfortunately adds some delay, which is noticeable on VoIP connections. In addition, the IP network could add some packet loss, and also some packet jitter due to some packets traveling on different routes and therefore not spending the same amount of time traveling across the network.

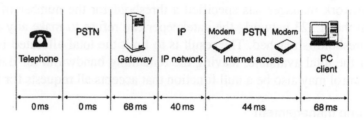

Delays shown are theoretical minimums

Figure 9.22
Factors introducing delays

9.6.3   Integrated services

Integrated services (int-serv) as described in RFC 1633 is a model developed by the Internet engineering task force. An end station that supports RSVP can request that bandwidth is reserved along a specific path, prior to the transmission of the data. RSVP has been designed to work in conjunction with routing protocols such as OSPF and BGP-4, and rely on these to decide where the reservation requests should be sent. There are four components in the int-serv model, namely RSVP, an admission control routine that determines if network resources are available, a classifier that puts packets in specific queues, and a packet scheduler that schedules packets to meet quality of service requirements.

9.6.4   Differentiated services

Differentiated services (diff-serv) as described in RFC 2474 was also developed by the IETF. It distinguishes packets that require different network services into different classes, identified by an 8-bit differentiated services (DS) field. This field then replaces the type of service field in the IPv4 header, or the traffic class field in the IPv6 header. It is assumed that Internet service providers will be able to offer different types of service, at different costs, based on the DS field.

9.6.5   Multiprotocol label switching

Multiprotocol label switching (MPLS) as described in RFC 2702. In a so-called MPLS domain, in which all routers are label switching routers (i.e. MPLS capable), all packets entering the domain are modified by inserting a 32-bit header ('tag') between the local network header and the IP header. Packets are classified at the ingress LSR and subsequently handled on the information contained in the tag and not the IP header. This means that this mechanism is independent of the network layer protocol being used.

9.6.6 Queuing and congestion avoidance mechanisms

Queuing mechanisms employed on routers classify incoming data flows according to attributes such as source and destination IP address, protocol used, or port number. It then lines them up in multiple parallel queues according to their classification. It then grants each of these individual flows a percentage of the available bandwidth.

Congestion mechanisms try to prevent data from being sent to already congested routes. This is done by using mechanisms such as random early detection (RED) to predict when congestion will occur, rather than leaving it up to TCP's built-in mechanism which causes packets to be dropped when congestion eventually does occur.

10

Cellular wireless communications

Objectives

After completing this chapter you should be able to explain the basics of the following:

- Cellular concepts such as cell structure, frequency reuse, cell splitting and call hand-over
- Analog cellular voice systems, in particular AMPS and N-AMPS
- Digital cellular voice systems, in particular D-AMPS (North American TDMA), CDMA and GSM
- Cellular data systems, in particular CDPD, GPRS and HDR
- 'Cordless' technologies, in particular CTS and DECT
- PCS
- WAP
- 3G mobile technologies.

10.1 Basic concepts

10.1.1 Mobile communication

Early mobile communications systems were fashioned in a way similar to television broadcasting systems. High-powered transmitters could service an area with a radius up to 50 km, and several of these could be deployed adjacent to each other. Each transmitter could communicate with several users at a time, using a pair of frequencies for each user. In order to provide access to the conventional PSTN, an operator would be available (Figure 10.1).

Interference problems caused by mobile units in adjacent areas showed that the same channels (frequencies) could not be used in adjacent areas. Areas had to be skipped before the same frequencies could be re-used. It was also discovered that the interference effects were not related to the distance between areas, but rather to the ratio of distance between areas to radii of the areas. By reducing the radius by half, the number of users per area could be increased by a factor of 2 squared, i.e. by four.

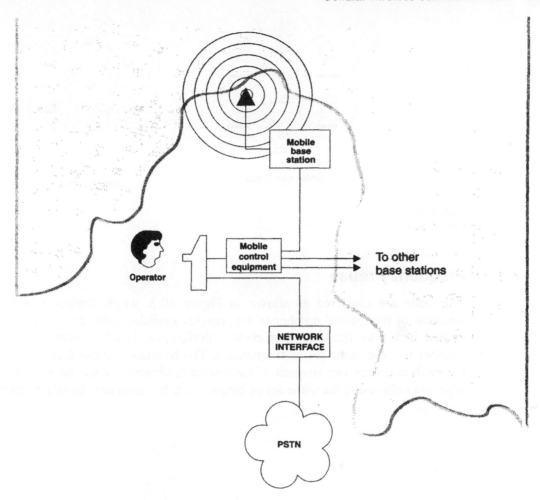

Figure 10.1
Basic mobile telephone network

10.1.2   Cellular communication

Cellular communications systems are wireless mobile communications systems that divide a large geographic area into smaller sections or cells, each with a low-power wireless transmitter, for the purpose of optimising the use of a limited number of frequencies. Out of a total of less than 1000 usable frequencies, a subset of frequencies are allocated to each cell and then re-used in several cells. Each mobile phone uses two frequencies, one from the phone to the cell transmitter and one from the cell transmitter to the phone. Several cells can simultaneously use the same subset of frequencies, providing they are not adjacent to each other as this will lead to interference.

The basic geographic unit of a cellular system is a cell that conceptually has a hexagonal (honeycomb) shape. In reality a cell cannot be hexagonal because of the radiation pattern of the antenna. Rather, it is round, but even this theoretically circular radiation pattern is modified because of geographical features such as mountains, hills, etc. (Figure 10.2).

In 1981 the FCC in the USA adopted rules creating a commercial cellular service and set aside 50 MHz of spectrum in the 800 MHz frequency band for two competing cellular systems in each market (25 MHz for each system). The 25 MHz assigned to each cellular system presently consists of 395 voice channels and 21 control channels each.

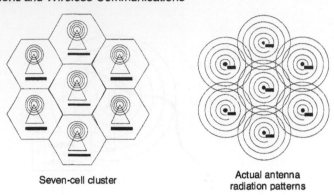

Seven-cell cluster

Actual antenna
radiation patterns

Figure 10.2
Cell cluster

10.1.3   Frequency reuse

The cells are clustered as shown in Figure 10.3, which depicts a seven-cell cluster.
Because of the limited number of frequencies available individual frequencies have to be
reused in a way that does not create interference. In this example, cells with a same
number use the same set of frequencies. The frequency re-use factor in this case is 1/7,
i.e. each cell uses one seventh of the available channels. It can be seen that there are no
adjacent cells using the same set of frequencies, hence there is no interference.

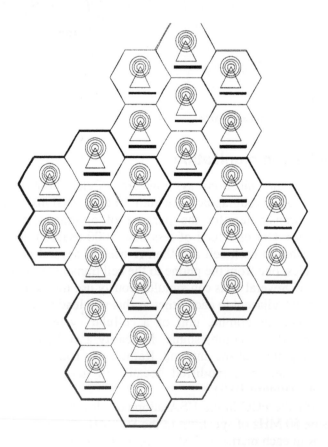

Figure 10.3
Frequency reuse

10.1.4   Cell splitting

As a cell becomes mature (full of users), the same approach can be used to split the cell into smaller cells. In this way urban cells with a high traffic density can be split into as many areas as necessary to provide acceptable level of service. As the cells are reduced in size, the transmitter power is reduced accordingly. This concept is illustrated in Figure 10.4.

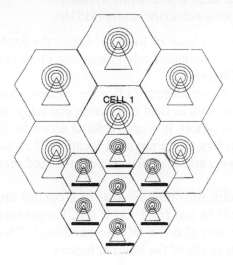

Figure 10.4
Cell splitting

10.1.5   Handoff

A problem is created when a mobile subscriber travels from one cell to another during a call. Since adjacent cells do not use the same channels (frequencies) a call must either be dropped or transferred from one channel to another when a user crosses the line between cells. Since dropping is unacceptable, a handoff process was created whereby the mobile telephone network automatically transfers the call from one channel to another as the mobile crosses over into an adjacent cell (Figure 10.5).

During a call, two parties are on one voice channel. When the mobile unit moves out of the coverage area of a given cell the reception becomes weak and the cell site in use requests a handoff. The system switches the call to a stronger channel on a new site without interrupting the call or alerting the user.

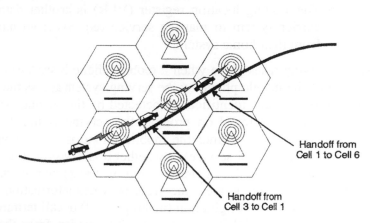

Handoff from
Cell 1 to Cell 6

Handoff from
Cell 3 to Cell 1

Figure 10.5
Handoff

10.1.6   Typical system components

A cellular communication system consists of the following four major components:

1. A public switched telephone network (PSTN)
2. A mobile telephone switching office (MTSO)
3. Cell sites with antenna systems
4. Mobile subscriber units (MSUs).

The PSTN is made up of local networks, the exchange area networks and the long-haul networks that interconnect telephones and other communication devices worldwide. It is discussed in Chapter 4.

The MTSO is the Central Office for mobile switching. It houses the mobile switching centre (MSC) as well as field monitoring and relay stations for switching calls from cell sites to PSTN central offices. In analog cellular networks, the MSC controls the system operation.

A cell site refers to the physical location of radio equipment servicing given cell. This includes a tower, antennas, RF transmitters and receivers, interface equipment and power sources.

A mobile subscriber unit consists of a control unit and a transceiver that transmits and receives signals to and from a cell site. Larger sets (4 W) are typically vehicle-mounted while smaller sets (0.6–1.6 W) are hand-held. The usage of the portable sets is limited due to the charge life of the internal battery.

10.1.7   Signaling

The signaling system used for mobile communication systems is SS7, which has already been discussed in detail in Section 4.8. There are two major types of intersystem signaling for mobile communication systems.

ANSI-41 is the standard used for TDMA and CDMA whilst GSM mobile application part (MAP) is the standard used for GSM. Despite differences, both systems have certain key things in common, in that they support MSCs, HLRs and VLRs as described below.

- A mobile switching center (MSC) is a telecommunication switch deployed in mobile communications networks to provide call control, call processing and call access to the PSTN.
- The home location register (HLR) is a database maintained by a users home carrier. The HLR stores user information such as user preferences and account status.
- The visiting location register (VLR) is another database used by the serving carrier system to manage service requests from mobile users who are away from their home system.

The following takes place when a mobile phone is switched on. Step one involves the detection of the mobile phone by the serving system across the air interface. Each mobile phone emits a unique identification detected by the serving system. The mere fact that a mobile phone is switched on prompts the serving system to do this. The database is queried to determine whether the mobile phone is in its 'home area' (the city or area where it is signed up for service) or whether it is a 'visitor' (i.e. roaming).

The second step involves determination of the appropriate handling of call requests. If the phone is in its home area, the HLR provides information regarding the handling of requests for either call origination (making a call) or call termination (being called). If the phone is roaming, a VLR must request information from the HLR so that the visited (serving) system can process calls appropriately.

These signaling and database communications typically occur before any call is received. By the time the user places or receives a call, the serving system already knows whether the account associated with the phone is in good standing and the location of the user so that calls placed to the user may be placed. Communication between the VLR in the serving system and the HLR of the home system is facilitated by mobile networking protocols and signaling based on SS7.

When a mobile phone places a call, the following takes place.

The phone scans the control channels since it needs to use the one with the strongest signal. To find the closest base station, it checks all 21 control channels and determines which one is strongest. It then chooses the stronger signal and decides to use that one for placing the call. Next, it sends an origination message (about 250 ms long) that contains the MIN (mobile identification number, the cell phone number (e.g. 082 404 6667), its ESN (electronic serial number) and the number just dialled. After the cellular service provider verifies that the caller is a valid paying customer, the base station sends a channel assignment message to the mobile phone informing it on which channel the communication takes place. The mobile phone tunes to the assigned channel and begins the call. The ring-back or busy signals heard are transmitted by the base station as audio signals.

When a call is placed to a mobile, the home switch reviews the HLR. If the mobile user is in the home area, the call is delivered. If the mobile user is roaming, the HLR indicates which VLR is currently maintaining the mobile user's records, after which the HLR uses SS7 and the appropriate mobile networking protocol to request delivery instructions from the VLR. The VLR then provides these instructions to the HLR allowing the home switch to deliver the call to the serving switch and ultimately to the mobile phone as if it were in the home area.

10.2   Cellular analog systems

10.2.1   Analog systems

Analog cellular operates in the 800 MHz frequency range and is available in 95% of the USA. Analog cellular transmits voice through the air using continues radio waves with Frequency Modulation (FM). The following list contains some of the systems currently in use. Of these, AMPS is by far the most popular.

Overview of existing systems

- AMPS was developed by Bell Labs and first used commercially in the USA in 1983. It is currently the world's largest cellular standard and operates in the 800 MHz band.
- C-450 was installed in South Africa during the 1980s. It was known as Motorphone and run by Vodacom SA. It used the 450 MHz band.
- C-Netz is an older cellular technology, similar to C-450, and used in Germany and Austria.
- Comvik was introduced in Sweden by the Comvik Network in 1981.
- TACS (total access communication system) was introduced in the UK in 1985 and operates in the 900 MHz frequency range. It has either 600 or 1000 channels, consisting of 42 control channels and either 558 or 958 voice channels.
- ITACS (International TACS) is a minor variation of TACS to allow operation outside of the UK by allowing some flexibility in assigning the control channels.

- ETACS (Extended TACS) is a current UK system (a replacement of TACS) which has 42 control channels and 1278 voice channels.
- IETACS (International ETACS) is, once again, a variation for use outside of the UK by allowing flexibility in assigning the ETACS control channels.
- NTACS (Narrowband TACS) has three times as many voice channels as ETACS, with the same speech quality.
- JTACS (Japan TACS) is a version of TACS for Japanese use.
- RC2000 (Radio Comm 2000) is a French system launched in 1985.
- NTT (Nippon Telegraph and Telephone) is the old Japanese analog standard.
- N-AMPS (Narrow Band Advanced Mobile Phone System) was developed by Motorola as an interim technology between analog and digital. It operates in the 800 MHz range and has three times greater capacity than AMPS.
- NMT-450 (Nordic Mobile Telephones/450) was developed by Ericsson and Nokia for service in the topologies of the Nordic countries (1981). It uses FDD FDMA at 450 MHz, with a range of 25 km.
- NMT-900 (Nordic Mobile Telephones/900) is a 900 MHz upgrade to NMT-450 to accommodate higher capacities.
- NMT-F is a French version of NMT-900.

Of all these standards, AMPS and its variations are the most widely deployed.

10.2.2   AMPS

Advanced Mobile Phone System (AMPS) is a first generation cellular technology that uses the 800–900 MHz frequency band with 30 kHz bandwidth for each channel. It was designed for urban use but later expanded to rural areas and was the first standardized cellular system used in the world. AMPS handsets have a telephone style user interface and are compatible with any AMPS base station, which makes roaming simpler for subscribers. It is used throughout the world, particularly in the USA, South America and China.

AMPS uses the frequency division multiple access (FDMA) technique to transmit information by dividing the available bandwidth into 30 kHz channels, and then transmitting an analog signal on each channel, using two channels per phone – one for incoming traffic and one for outgoing traffic. The concept of FDMA is illustrated in Figure 10.6.

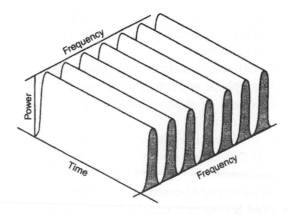

Figure 10.6
FDMA concept as used by AMPS

AMPS, unfortunately, suffers from a few drawbacks. With only one signal per channel, the system has a limited call capacity. The available frequency spectrum is limited and there is no room for spectrum growth, hence the call capacity of cells cannot be extended by increasing the spectrum. The system is not geared towards data communications, it is easy to eavesdrop on conversations and there is inadequate protection against fraud.

NAMPS

Narrow band analog mobile phone service is a second-generation analog cellular system, designed as an upgrade to AMPS. Its major aim is to solve the problem of low calling capacity. It combines existing voice processing with digital signaling, tripling the capacity of existing AMPS systems. NAMPS uses FDM to fit three channels into the AMPS 30 kHz channel bandwidth, thereby increasing the capacity by a factor of three. Unfortunately, this increases the possibility of interference because channel bandwidth is reduced.

10.3   Digital systems

The calling patterns for the PSTN do not apply to mobile systems. Where the average PSTN phone calls may last more than 10 min, mobile calls are often less than 90 s. As a result, mobile phone users on early systems often found that they could not get dialtone due to insufficient circuits and/or equipment. This is known as call-blocking probability. As a result, systems quickly became saturated and the quality of service decreased. In order to increase capacity, there was a move to digital systems such as time division multiple access (TDMA), global system for mobile communications (GSM), personal communication service (PCS1900) and code division multiple access (CDMA) as these significantly increase the efficiency of cellular systems and allow a greater number of simultaneous conversations. Technologies such as TDMA and CDMA not only offer more channels in the same bandwidth, but also allow encrypted voice and data.

10.3.1   Overview of existing systems

- A1-Net is the Austrian name for GSM 900 networks.
- B-CDMA (broadband CDMA), now known as W-CDMA, was developed by Ericsson and is to be used in UMTS.
- CDMA (code division multiple access). This will be discussed in more detail.
- cdmaOne is another name for first generation CDMA (IS-95) mentioned above.
- cdma2000 is the new second-generation CDMA specification for inclusion in UMTS.
- CT-2 is a second-generation digital cordless standard, which offers 40 voice channels.
- CT-3 is a third-generation digital cordless telephone, very similar to and a precursor to DECT.
- DECT (digital european cordless telephone) uses TDMA with 12 time slots. It will be discussed in more detail in this section.
- CTS is a GSM cordless telephone system. This will also be discussed in more detail.
- D-AMPS (IS-54). Digital AMPS is a variation of AMPS that uses three timeslot TDMA channels.
- DCS-1800 is a digital cordless standard (1992), now known as GSM 1800. This is a different version of GSM, and 900 MHz phones cannot be used on DCS 1800 networks unless they are dual band.

- EDGE, known as UWC-136 (enhanced data rate for GSM evolution) will allow GSM operators to use existing radio bands to offer wireless mullet-media IP-based services.
- E-Netz is the German name for GSM 1800 networks.
- GERAN is used to describe GSM and EDGE based on 200 kHz radio access network.
- GMSS is the geostationary mobile satellite standard, a satellite air interface developed from GSM.
- GSM (global system for mobile communications) was the first European digital standard (1991) and over 80 GSM networks are now in operation world wide. It operates at 900 MHz.
- IDEN (integrated digital enhanced network) is a private mobile radio system by Motorola operating in the 800 MHz, 900 MHz and 1.5 GHz bands. It uses TDMA.
- iMode is a system from NTT DoCoMo that uses compact HTML to provide WAP-like content to iMode phones.
- IMT DS wideband CDMA, or WCDMA.
- MT MC is also known as cdma2000.
- IMT SC is another designation for UWC-136, commonly known as EDGE.
- IMTFT is better known as DECT.
- Inmarsat is an international maritime satellite system using a number of GEO satellites.
- Iridium is a mobile satellite phone/pager network using TDMA for inter-satellite links (2 GHz band).
- N-CDMA (narrow band code division multiple access) or plain old original CDMA, known in the USA as IS-65.
- PCS (personal communication service) operates in the frequency band 1050–1990 MHz (1995). This encompasses a wide range of new cellular standards such as N-CDMA and GSM1900.
- TDMA (time division multiple access) was the first USA digital standard to be developed. To be discussed in more detail.
- TETRA (terrestrial trunked radio) is a new digital trunked radio standard defined by the European telecommunications standards institute (ETSI) developed for professional mobile radio users.
- UMTS (universal mobile telephone standard) is a next generation global cellular system, which should be in place by the year 2004. It uses a combination of TDMA and W-CDMA operating at around 2 GHz with proposed data rates of up to 2 Mbps.
- W-CDMA is one of the components of UMTS along with TDMA and CDMA2000. It has an air interface of 5 MHz.
- WLL (widest local loop) is used to connect subscribers in remote areas where fixed lines are impossible. Most modern WLL systems use CDMA.

10.3.2   TDMA

Time division multiple access (TDMA) is a digital transmission technology that allows several users to access a single radio frequency channel without interference by allocating timeslots to each user within its channel. TDMA multiplexes three signals over a single channel thereby providing a 3 to 1 gain in capacity over analog cellular (AMPS).

TDMA is basically the frequency division multiple access (FDMA) technique used by AMPS with a timesharing facility built into the system, allowing three users to share each channel (Figure 10.7). The process works as follows: assuming the voice signals of three users A, B and C are to be transmitted over a specific frequency channel. Time-wise, each channel is cyclically divided into 6 ms frames, with each frame subdivided into six equal timeslots of 1 ms each. Each timeslot carries a header, together with digital data representing a speech sample. Digitized samples of voice A will be fitted into timeslots 1 and 4, followed by a sample of Voice B in slots 2 and 5, and samples of voice C in slots 3 and 6. The whole process is then repeated. Effectively, the IS-54 implementation of TDMA immediately triples the capacity of analog cellular frequencies by time-division multiplexing each 30 kHz channel between three users. It is predicted that with intelligent antennas, adaptive channel allocation and other techniques the capacity could approach 40 times analog capacity.

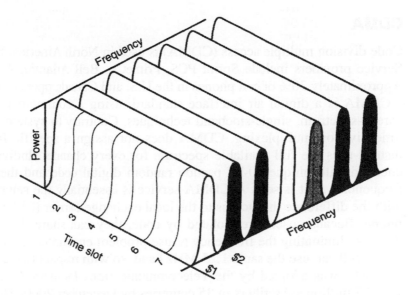

Figure 10.7
TDMA concept

TDMA has a weakness in that once a particular time slot is allocated to a specific user, it often carries no information due to pauses in the conversation. This problem is addressed by an enhanced version of TDMA (E-TDMA) that effectively performs statistical multiplexing. When subscribers have something to transmit, a bit is inserted in a buffer queue. When that is detected by the system, it allocates bandwidth to that source accordingly. If a subscriber has nothing to transmit, the system detects it through the absence of the bit and the empty slot goes to the next subscriber so that the particular timeslot is not wasted.

TDMA advantages include the following:

- It can be easily adapted to carry data as well as voice and has the ability to carry data rates from 64 kbps up to 120 Mbps, expandable in multiples of 64 kbps. This simplifies transmission of fax and SMS services.
- The cell phone is only transmitting for 1/3–1/10 of the duration of the conversation, which extends battery life.
- TDMA separates users in time, which means that they do not experience interference due to simultaneous transmissions.

- TDMA base station equipment is significantly cheaper than CDMA equipment.
- TDMA is the most effective technology for upgrading a current analog (AMPS) system to digital.

The technology does, unfortunately, have some disadvantages as well.

- Although users may have been allocated timeslots, users roaming from one cell to another may find that all the slots in the next set are already occupied, leading to a dropped call. If all timeslots in a cell are already occupied, a user will also not receive a dial tone.
- Another problem with TDMA is that it is subject to multi-path distortion due to different path lengths between directly received signals and reflected signals. This problem can be addressed by putting time limits on signals and ignoring signals that arrive after a certain time.

10.3.3   CDMA

Code division multiple access (CDMA) is used in North America, Korea and parts of Asia. Service providers include Sprint PCS, PrimeCo, Bell Atlantic, AirTouch and Ameritech. Approximately 30% of cell phones in the U.S. and Canada operate on CDMA.

CDMA is a digital air interface standard, using a commercial adaptation of military, spread-spectrum, single-sideband techniques. Contrary to systems such as GSM that use time division multiplexing, CDMA does not assign a specific frequency to a user but instead uses the full available spectrum for every channel. Individual conversations are encoded with unique 40-bit pseudo random digital code and then spread over multiple frequencies as it is sent. A CDMA service is essentially the same as a wire-line service, with the difference that access to the local exchange carrier (LEC) is provided by wireless phone. Because users are isolated by code, they can share the same carrier frequency, thereby eliminating the frequency re-use problem encountered with AMPS and DAMPS. Every cell can use the same 1.25 MHz band, so with respect to clusters n = 1.

CDMA was adopted by the Telecommunications Industry Association (TIA) in 1993, with 27 million subscribers in 35 countries by December 2000. There are also 43 wireless local loop (WLL) systems in 22 countries using CDMA technology.

There are currently a number of 'flavors' of CDMA.

- CDMA (also known as first generation narrowband CDMA, N-CDMA or IS-95) competes with GSM technology. CDMA is characterized by a high capacity and small cell radius, employing spread spectrum technology and a special coding scheme.
- A variation of IS-95, named IS-95B, provides ISDN rates up to 64 kbps.
- W-CDMA is a variation of CDMA developed by Ericsson, touted as a possible third generation (3G) technology as opposed to Qualcomm's cdma2000.
- CdmaOne is just another name for the original CDMA, above.
- Cdma2000 is a new 3G spec for inclusion in UMTS, and is opposition for W-CDMA. Cdma2000 is implemented in phases.

Phase 1 of cdma2000, which is now available, includes an enhancement of cdmaOne known as 1XRTT that enables a 144 kbps packet data rate in a mobile environment. Battery standby time and voice capacity was also increased by a factor of two over the original specifications and all these capabilities are available in an existing cdmaOne 1.25 MHz channel. It also transmits identifying information from the handset, enabling emergency service providers to locate its position.

Phase 2 of cdma2000 focuses on so-called third-generation (3G) CDMA technology specified in cdma2000. This includes multi-Carrier cdma2000 $1 \times MC$ and cdma2000 $3 \times MC$):

- cdma2000 $1 \times MC$ ($1 \times$ multi-carrier) is a CDMA mode in the International Telecommunications Union IMT 2000 family of standards which offers third-generation capabilities within a single 1.25 MHz channel. cdma2000 doubles the voice capacity of existing CDMA systems and offers data speeds of up to 7 kbps.
- cdma2000 $3 \times MC$ ($3 \times$ multi-carrier) is a CDMA mode in the ITU IMT 2000 family of standards which utilizes three standard 1.25 MHz channels within a 5 MHz band, offering data speeds up to 2 Mbps.

Whilst Qualcomm is promoting cdma2000, the developers of W-CDMA are pushing WCDMA/UMTS/direct spread, yet another mode in the IMT 2000 family of standards. It utilizes one 5 MHz channel, offering data speeds of up to 2 Mbps. It proposes a faster chip than cdma2000, but uses the spectrum less efficiently and is not as compatible with earlier versions of CDMA as cdma2000.

CDMA claims several advantages including:

- High quality voice because of the variable rate voice encoder/decoder (vocoder) used which eliminates background noise and cross-talk.
- A data capability due to the fact that CDMA networks are built around standard IP packet data protocols.
- Privacy due to the encoded spread spectrum transmissions designed with about 4.4 trillion codes which makes cloning and eavesdropping almost impossible.
- It is claimed that CDMA systems can provide 10–20 times the capacity of analog systems and more than three times the capacity of other digital systems in the same bandwidth.
- No frequency re-use problems.
- CDMA handsets use low amounts of power, which is significant in the light of fears about cell phones causing cancer.
- Calls are handed off through a 'soft' handoff method that is superior to TDMA and AMPS handoff, resulting in fewer lost calls during handoff.

CDMA does, however, have a drawback. There is no 'hard' limit on the number of users per cell. However, each user, particularly one transmitting with excessive power, adds to the overall noise within the system. Precise power control of mobiles is critical. Each mobile should transmit at the absolute minimum power necessary to ensure acceptable service quality.

10.3.4   GSM

GSM, originally groupe speciale mobile, but also known as global system for mobile communications, is a standard set by the European Union. It is currently used in Europe, the Far East, much of South America, New Zealand, Australia and South Africa.

GSM operates in the 900 MHz range, using narrowband TDMA, which allows eight simultaneous calls on the same radio frequency. The duplex distance, i.e. the distance between the uplink and downlink frequencies, is 80 MHz with a channel separation of 200 kHz and an over-the-air bit rate of 270 kbps. GSM was first introduced in 1991 and by the end of 1997 it was available in more than 100 countries, becoming the de facto standard in Europe and Asia.

The GSM specification defines the functions and interface requirements for GSM systems, but not the exact details of the hardware, in order to be as flexible as possible for various vendors. A GSM network consists of three subsystems namely the switching system (SS), the base station system (BSS) and the operation and support system (OSS) as shown in Figure 10.8.

The switching system (SS) is responsible for call processing and subscriber-related functions. It includes the following functional units:

- Home location register (HLR)
- Mobile services switching center (MSSC)
- Visitor location register (VLR)
- Authentication center (AUC) and
- Equipment identity register (EIR).

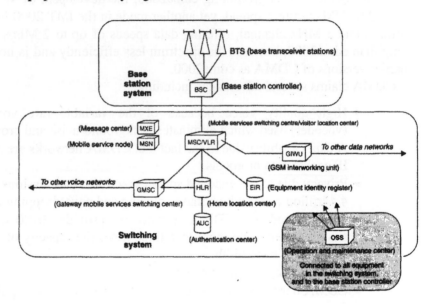

Figure 10.8
GSM subsystems

The base station system performs all radio-related functions and comprises two subsystems namely the base station controllers (BSCs) and the base station transceiver stations (BTSs). The BSCs are high-capacity switches that provide all the control functions and physical links between BSCs and BTSs such as hand-over, cell configuration and radio frequency power control. The BTSs, in turn, comprise the radio equipment such as transceivers and antennas needed for each cell, and thus handle the radio interface to the mobile stations. Several BTSs are controlled by a single BSC.

The operations and maintenance center (OMC) is connected to all equipments in the switching system and to the BSC. Its purpose is to offer support of centralized, regional and local operational and maintenance activities required for the network.

In addition to these basic functions, the system can also contain the following functional units:

- Message centers (MXEs) that provide integrated voice, fax and data messaging.
- Mobile service nodes (MSNs) that handle the mobile intelligent network (IN) services.

Gateway mobile services switching centers (GMSCs) that are nodes used to interconnect two networks. This function is often implemented in a MSC.

GSM internetworking units (GIWUs) that consist of hardware and software providing an interface to various other data communications systems. This allows users to alternate between data and switch during the same call.

Network areas

The GSM network consists of geographic areas that include cells, location areas (LAs), mobile services switching center/visitor location register (MSC/VLR) service areas and public land mobile network (PLMNs) (Figure 10.9). These concepts will now be discussed individually.

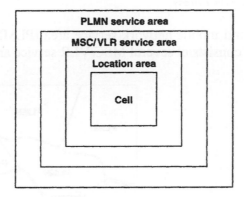

Figure 10.9
GSM network areas

A cell is the area covered by one base transceiver station and a cell global identity (CGI) number identifies it.

A location area, in turn, is a group of cells, each served by one or more base station controllers, but by a single MSC. Each LA is identified by a location area identity (LAI) number (Figure 10.10).

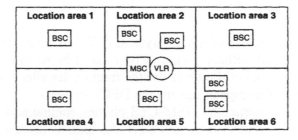

BSC: Base station control
MSC: Mobile services switching center
VLR: Visitor location register

Figure 10.10
GSM location areas

An MSC/VLR service area is a part of the network controlled by one MSC and is registered in the VLR of the MSC (Figure 10.11).

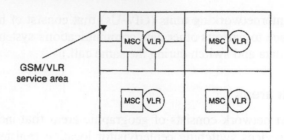

MSC: Mobile services switching center
VLR: Visitor location register

Figure 10.11
GSM MSC/VLR service areas

A public land mobile network service area (PLMN) is an area served by one network operator and consists of several MSC/VLR service areas (Figure 10.12).

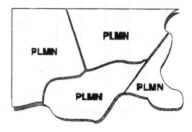

Figure 10.12
GSM PLMN areas

As is the case with conventional PSTN services, GSM provides a range of subscriber services including dual-tone multi-frequency (DTMF), CCITT Group 3 facsimile, short message service (SMS), cell broadcasts (short messages to all subscribers in a geographical area, voice mail, fax mail, call forwarding, barring of incoming and outgoing calls, advice of charge services, call hold, call waiting, conference calls, calling line identification and closed user groups.

10.3.5   CTS

The GSM cordless telephony system (CTS) is a new feature of the GSM standard that allows a GSM cellular phone to be used at home in conjunction with a fixed network. CTS calls via a GSM mobile use the frequencies allocated to E-GSM 900 MHz and GSM 1800 MHz. It competes against the DECT system.

Within the home, the GSM-CTS phones communicate with a CTS home base station HBS that is connected to the PSTN. As soon as the cellphone moves out of range of the HBS, it communicates via the normal cellphone network. No modifications to the cellphone is required other than a software upgrade.

10.3.6   DECT

Digital enhanced cordless telecommunications (DECT) is a digital wireless technology that originated in Europe and is now being adopted worldwide for cordless telephones, wireless offices and wireless telephone lines to homes (wireless local loop applications). It is related to GSM, but it is by contrast only a radio access technology rather than a comprehensive system architecture. It can internetwork with many other network types such as PSTN, ISDN and GSM networks.

DECT is a standard created by the European Telecommunications Standards Institute (ETSI) out of the UK CT-2 standard and the Swedish CT-3 standard, and defines a protocol for secure mobile communications. It uses TDMA technology to provide ten 1.75 MHz channels between 1.88 and 1.90 GHz, with each channel capable of carrying 12 simultaneous two-way conversations. DECT users can make and receive calls when in range of a radio base station over a distance of 20–50 m indoors, and up to 300 m outdoors, with a speech quality comparable with conventional wire-line phones.

The principal applications for the DECT standard include:

- Mullet-cell cordless communication systems for business applications. This is basically a DECT extension to a PBX.
- DECT/GSM (dual-mode) phones. Work is underway to combine DECT cordless networks within GSM cellular networks. The DECT/GSM interworking profile enables a handset to access both networks so that users can make use of the virtually free PBX service within a corporate facility and then switch over to the GSM cellular network when the handset is out of range of the base station.
- Single-cell home cordless phones.
- Radio access systems for subscribers to public telecom services where wire-line services are difficult or uneconomical to install, for example in rural areas (wireless local loop applications). The subscriber uses a standard telephone and does not have any communications mobility. However, if subscribers can be given a cordless phone to provide limited mobility, the service is referred to as cordless terminal mobility (CTM).

A big advantage of DECT is dynamic reconfiguration which means that no advance load, frequency or cell planning has to be done for a DECT implementation. In a conventional cellular networks, frequency reuse has to be carefully planned to eliminate cross interference.

With DECT, the 12 conversations that are occurring at any one time can take place on any of the ten channels in any combination. Any handset can identify any open frequency and time slot on the nearest base station and grab it. Therefore adding a base station requires no modification of existing base stations and no prior planning of channel allocations. Compared to conventional analog systems DECT does not suffer from interference or cross talk since DECT dynamically reconfigures itself to make the best use of available frequencies and time slots.

Voice compression and the higher levels of the DECT protocol are not implemented at the base stations but are handled separately by a concentrator that is the interface between the wireless network and the PSTN.

10.4 Wireless data systems

10.4.1 CDPD

One of the alternatives for sending data over an analog mobile network such as AMPS is to dial up a connection and send the data across with a modem at each end, using appropriate protocols. Since the method does not involve any packet switching, but merely creates a circuit-switched connection between the two end points, this method is referred to as circuit switched cellular (CSC).

A more elegant, robust and secure method is, however, cellular digital packet data or CDPD, also known as 'wireless IP'. CDPD is a data transmission technology for use on analog cellular phone frequencies, specifically the unused cellular channels in the 800–900 MHz range. Although CDPD piggybacks on top of the cellular voice infrastructure it is not constrained by the 3 kHz limit on voice, but rather uses the entire 30 kHz bandwidth of the channel. Information is sent over the air at a data rate of 19.2 kbps. Because of protocol overheads, the actual data throughput is somewhat less, in the vicinity of 12–14 kbps. CDPD features high reliability and security, and seamless roaming at a reasonable cost and with no interference with voice traffic.

In accordance with the Internet protocol (IP) data is packaged into discrete packets of information for transmission over the CDPD network. In addition to IP addressing information, each packet includes the necessary information (fragmentation flags and fragmentation offset) to enable the receiver to reassemble the data at the receiving end.

Analog voice signals originating from mobile phones are transmitted to and from cells (as shown in Figure 10.13) to a mobile base station (MBS) using frequency modulation (FM). From the MBS, voice traffic is then routed through the mobile telephone switching office (MTSO) and then over the public switched telephone network (PSTN).

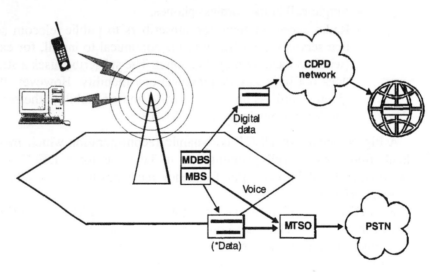

Figure 10.13
CDPD voice and data communication paths

CDPD operates differently. Digital data packets are sent over the same frequency spectrum as analog voice, but with different modulation in the air interface. CDPD calls use AMPS channels that are not being used for voice calls. It either uses whatever is available or, alternatively, carriers may assign one or more dedicated channels to CDPD. The data packets travel from the mobile to the cell where it is received by the mobile data base station (MDBS). The MDBS then sends the data to the nearest mobile data intermediate system (MD-IS) which sends it across the CDPD network.

Communicating over CDPD is typically a client server application, therefore there needs to be a connection from the CDPD network to a fixed corporate network where the server resides. This can be done via the Internet, a leased line, or a frame relay network. At the corporate network site, access to the network is via a conventional IP router (Figure 10.14).

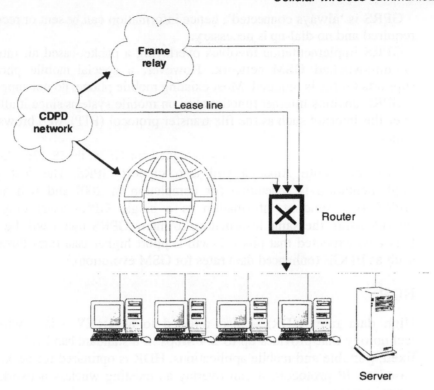

Figure 10.14
CDPD backend connections

Security is accomplished as follows:

- Firstly, an encrypted channel has to be established
- Secondly, the mobile device has to be authenticated and
- Thirdly the data has to be encrypted for transmission over the air link.

To accomplish the first two, the mobile client and the home MD-IS server exchange secret encryption keys for decoding the data, along with dynamically changing authentication codes for authorising the mobile device. This is done by means of Diffie-Hellman public key encryption negotiation. The third task is accomplished through the use of RSA data security's RC-4 encryption standard.

CDPD charges are not based on call duration but rather on the amount of data transferred which is roughly US$0.05 per kilobyte, together with a relatively small monthly service fee. It is ideal for e-mail, Internet access, point of sale terminals, taxis, remote alarm monitoring, telemetry, or any application requiring access to a remote network or database. An added advantage is the fact that it is 'always on' and no dial-up is necessary before exchanging data.

10.4.2   GPRS

General packet radio service (GPRS) is a data service (non-voice) that supplements circuit switch data (CSD) and short message service (SMS) on GSM networks. It can theoretically run at speeds of up to 171.2 kbps, achievable by using all the eight time slots on a GSM network. In practice the maximum speed is about 150 kbps. This is in contrast with circuit switched data that runs at 9.6 kbps. Because of its bandwidth, GPRS suitable for sending and receiving small bursts of data such as e-mail, or downloading larger files e.g. when web browsing.

GPRS is 'always connected', hence information can be sent or received immediately as required and no dial-up is necessary.

GPRS implementation involves overlaying a packet-based air interface on an existing circuit-switched GSM network. However, a special mobile phone or terminal that supports GPRS is required. Most existing mobile phones do not support GPRS.

GPRS enables Internet functionality on mobile systems since it allows all services used over the Internet such as the file transfer protocol (FTP), web browsing, chat, e-mail and telnet.

Although GPRS was originally designed for GSM, the IS-136 TDMA standard used in North and South America will also support GPRS. The first phase of the GPRS implementation is scheduled for completion in 2001 and will support point-to-point GPRS (i.e. sending information to a single GPRS user) only. Point-to-multipoint (broadcasting the same information to many GPRS users) will be included in phase 2. It is also expected that phase 2 will support higher data rates through using techniques such as EDGE (enhanced data rates for GSM evolution).

10.4.3   HDR

High data rate (HDR) – also referred to as 1× EV – is a wireless Internet access technology that provides up to 2.4 Mbps in a standard bandwidth 1.25 MHz channel for fixed, portable and mobile applications. HDR is optimized for packet data services and is based on IP protocols. It can overlay an existing wireless network or work as a stand-alone system.

The Telecommunications Industry Association (TIA) recently adopted a specification based on HDR, known as TIA/EIA/IS-856 'CDMA-2000, High Rate Packet Data Air Interface Specification', also known as 1× EV. This specification was developed by the Third Generation Partnership Project 2 (3GPP2, a partnership consisting of 5 telecommunications standards bodies: CWTS in China, ARIB and TDC in Japan, TTA in Korea and TIA in North America).

1×EV technology can be embedded in handsets, laptops and notebooks, and supports e-mail, web browsing, e-commerce and many other applications. It offers up to 2.4 Mbps air link in a 1.25 MHz channel, packet data capability, IP compatibility and compatibility with existing CDMA systems.

10.5   PCS

The TIA IS-136 specification defines a personal communications service (PCS) that uses the IS-54 digital mobile system (a.k.a. D-AMPS or North American TDMA) as a basis. The core of IS-136 is the digital control channel (DCCH) which is an enhancement to the existing TDMA technology. The IS-136 air interface comprises analog voice channels (AVCs), analog control channels (ACCSs) and digital traffic channels (DTCs). Regardless of the total number of channels, IS-136 takes ONE of the DCCs and converts it to a digital control channel or DCCH. Figure 10.15 gives a very simplified view of this arrangement. It shows a hypothetical system with one of the DTCs replaced with a DCCH.

The information carried on the DCCH flows in two directions over the air interface from phone to system (uplink) and from system to phone (downlink). PCS phones and other DCCH-capable equipments continuously monitor ('camp on') the DCCH channel in each sector of an IS-136 system. They receive control information over this channel, switch to an allocated channel to send or receive information, and then return to the DCCH.

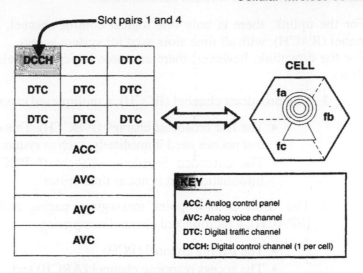

Figure 10.15
IS-136 DCCH

Whereas the cellular (voice) system uses the band from 824 to 849 MHz for phone transmit (832 channels of 30 kHz) and the band from 869 to 894 MHz for base transmit (832 channels). PCS uses the data band from 1850 to 1910 MHz for phone transmit (1841 channels of 30 kHz) and from 1930 to 1990 MHz for base transmit (1841 channels of 30 kHz).

The IS-136 air-interface model contains four layers, as shown in Figure 10.16.

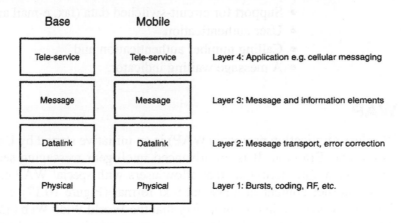

Figure 10.16
1s-136 Air-interface model

The physical layer (layer 1) deals with the physical radio interface, bursts, slots, frames, etc.

The data link layer (layer 2) deals with data packaging, error detection, error correction, etc.

The message layer (layer 3) creates and handles the messages sent and received through the air.

The application layer (layers 4–7) handles the current teleservice being used, for example messaging services.

The physical control channel (DCCH) described here is subdivided into several logical Channels.

For the uplink, there is only one logical control channel, namely the random access channel (RACH), with all time slots used for system access.

For the downlink, however, there are three logical channels, once more subdivided as follows:

1. The broadcast channel (BCCH), a multiplexed channel comprising:

   - The fast broadcast channel (F-BCCH). This channel handles information that phones need immediately, such as system ID information and
   - The extended broadcast channel (E-BCCH). This channel handles information that is not as time critical.

2. The SMS point-to-point messaging, paging and access response channel (SPACH), a multiplexed channel comprising:

   - The paging channel (PCH)
   - The access response channel (ARCH) and
   - The SMS point-to-point messaging channel (AMSCH).

3. The shared channel feedback (SCF), which provides a collision-free mechanism for the uplink. PCS features include:

   - Increased resistance to eavesdropping
   - SMS facilities
   - A sleep mode for extending handset battery life
   - Support for international roaming because of the dual-frequency phones used
   - Support for circuit-switched data (fax, e-mail and Internet access)
   - User authentication
   - Calling number authentication and
   - A message waiting indicator.

10.6   WAP

Wireless application protocol (WAP) is an initiative started by Unwired Planet, Motorola, Nokia and Ericsson. It is an advanced intelligent messaging service for mobile phones and other mobile terminals that allow users with special WAP-enabled mobile phones to view Internet content in a special text format (Figure 10.17). It hides the complexity of the cellular system in the same way that the World Wide Web hides the complexity of the Internet from its users. WAP supports most wireless networks including CDMA, GSM and TDMA. It is also supported by many operating systems, especially ones engineered for handheld devices such as PalmOS, Windows CE, FlexOS, OS/9 and JavaOS. Users can surf the Internet in text format, send and receive e-mail, buy and sell stocks, book theatre tickets, get traffic directions and order flowers, to name but a few applications.

WAP defines the following:

- A thin-client 'microbrowser' using the WML standard that is optimized for handheld devices. WML (wireless markup language) is a stripped down version of the hypertext markup language (HTML) used for creating Internet web pages. WML is scaleable from one-line text displays up to complex graphic screens.
- A proxy server that acts as a gateway between the wireless network and the Internet, providing data transfer and protocol translation for the handset.

- A computer-telephony integration (CTI) applications programing interface (API) called WTAI, between data and voice. This enables the handheld device to operate as both a terminal and a phone and allows manufacturers and content developers to develop their own applications and services.

Figure 10.17
WAP phone display

The WAP phone also has on-board memory that can be used for address books, text input, bookmarks, etc.

WAP utilizes Internet standards such as eXtensible markup language (XML), User datagram protocol (UDP) and Internet protocol (IP). Many of the protocols are based on Internet standards such as HTTP but had to be optimized since these protocols are inefficient over mobile networks where large amounts of mainly text based data has to be sent.

WAP is designed to operate on several wireless networks such as GSM-900, 1800 and 1900 MHz, IS-136 (North American TDMA), DECT, TDMA, PCS, FLEX and CDMA. All network technologies and bearers will also be supported including short message service (SMS), circuit switched cellular data (CSD), cellular digital packet data (CDPD) and general packet radio service (GPRS).

Figure 10.18 illustrates the architecture of a WAP gateway.

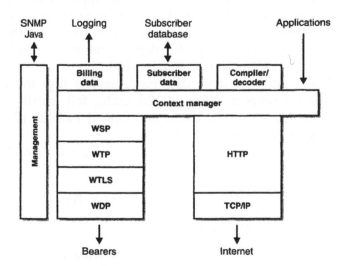

Figure 10.18
WAP gateway architecture

The following protocols are used on the WAP side:

- WDP (WAP datagram protocol) is the transport layer that sends and receives messages via any available bearer network such as SMS or CSD.
- WTLS (wireless transport layer security) is an optional security layer with encryption facilities as required for e-commerce transactions.
- WTP (WAP transaction protocol) provides transaction report adding reliability to the datagram service provided by WDP.
- WSP (WAP session protocol) provides a session layer to allow exchange of data between applications.

On the Internet side, the following protocols are used:

- HTTP (hypertext transfer protocol) retrieves WAP content from the Internet as requested by the mobile device.
- TCP (transmission control protocol) provides the end-to-end control.
- IP (Internet protocol) provides the routing and fragmentation services required.

10.7   3G systems

3G is an ITU specification for a 'Third Generation' mobile communications technology to be deployed in a critical mass commercially by the years 2004. The first analog systems (such as AMPS) are considered 'First Generation' while the present digital systems are 'Second Generation'. 3G terminals will be more complex than current GSM handsets due to the requirement for video, additional storage, multiple modes and a longer battery life.

One of the ideas behind the 3G movement is to create one global standard as opposed to the current fragmented approach. In December 1998 the third generation partnership project (3GPP) was created after an agreement between six international standards bodies including ETSI, ARIB and TIC from Japan, ANSI of the USA and TTA of Korea. The Chinese and the CDMA development group were not original members of 3GPP. The 3GPP's brief is to prepare, approve and maintain the technical specifications and reports for a third-generation mobile system. The proposed 3G standard is known as IMT-2000 and describes a CDMA-based standard that encompasses three optional modes of operation, each of which should work over GSM MAP and IS-41 networks.

The first proposed mode is W-CDMA, supported by Japan's ARIB as well as GSM network operators and vendors for deployment in Japan and Europe. The second proposed mode is cdma2000, supported by cdmaOne operators and members of the CDMA development group, for deployment in the USA. The third mode is UTRA (3G terrestrial radio access) favored by China.

IMT-2000 proposes three data rates, namely high mobility for outdoor mobile use (144 kbps for users traveling at 120 km/h), full mobility (384 kbps for outdoor users traveling at less than 120 km/h) and limited mobility (2 Mbps for stationary or slow moving users at less than 2 km/h).

Interested readers can get more information at www.mobile3G.com

10.8   Terrestrial trunked radio (TETRA)

10.8.1   Private mobile radio systems

Over the last 10 years there has been a gradual evaluation of radio communications towards digital modulation techniques. This started with point-to-point high-capacity microwave radio communications, operating in the frequency bands above 1 GHz. Then point-to-multipoint microwave systems were developed. The first half of the 1990s saw the development of digital cellular technology. This technology is now being implemented as standard practice around the world for cellular mobile telephone communications.

Private mobile radio (PMR) technology has been a lot slower in the uptake of digital radio technology. Although several vendors have advertised products that are supposedly digital in nature, until very recently there has been little acceptance of these products. Also during the early part of the 1990s a number of proprietary standards and products were developed that were not well accepted. There are several standards currently under development which are being proposed as international standards.

There are currently two main standards under development worldwide:

1. APCO-25
2. TETRA.

The first standard is coming out of America. This is being developed by a very influential public safety lobby group referred to as APCO. They officially do not release standards, but what they refer to as *recommendations*. Their first well known radio recommendation in the area of public safety was APCO-16. This was a user recommendation for operational procedures of trunked radio systems by public health and safety groups such as the police, fire brigades, ambulance and strategic armed forces. This standard contained very little technical detail.

Around the late 1980s APCO started work on a technical and operational standard for a digital modulation radio system for both trunked and single-channel PMR. The standard is referred to as APCO-25. This standard was ratified in August of 1995. Manufacturers are now seriously setting about developing chips, hardware and software for operation using this standard.

The system uses digital signal processing to simulate the human voice and then only transmits the bits that represent a change in the human voice pattern, rather than the complete data description of the human voice (using straight eight-bit resolution PCM, for example). The software and microchips that carry out the processing of the human voice are referred to as *vocoders*. APCO-25 transmits over 12½ kHz bandwidth channels using a 9.6 kbps data rate (that is 0.77 bits per hertz). The modulation technique used here is referred to as C4 FM, which is a type of continuously shifted quadrature phase shift keying (QPSK).

APCO-25 was designed fundamentally to service the public safety service industry. It therefore operates like a standard two-way radio system in nature. This is essential for the quick radio access times required in this industry. Other industries that would be well served by radio equipment using this standard are the mining and manufacturing industries and to some extent, the utilities.

Data transfer capabilities are also included in the standard. The standard also includes a lot of strict requirements for intrinsic safety and weather protection. It is also of interest to note that the digital signal processing standard recommended for use in the vocoder is designed to provide the least amount of distortion to the human voice when there is a high level of background noise such as helicopters, police and fire engine sirens, gun shots, etc.

10.8.2   TETRA PMR standard

The TETRA standard is being developed by the European Telecommunications Standards Institute (ETSI), which established a memorandum of understanding (MOU) in 1994, and now has 56 members across 19 countries. Although the standard has been accepted in concept by a lot of European countries, there are a number of influential countries who are not totally satisfied with the present standard format and who are proposing a number of substantial changes.

TETRA operates in the 380–400 MHz, 410–430 MHz and 450–470 MHz bands. It requires only 25 kHz per carrier, as opposed to the 200 kHz for GSM and can co-exist with analog systems on adjacent channels.

TETRA is designed more for the European PMR market where commercial applications are considered as being of more importance. The standard is designed to service a lot of users who have commercial voice requirements and a heavy data transfer requirement.

The standard is based on time division multiplexing (TDM) technology, where four time slots are allocated to one 25 kHz bandwidth channel. The data transmission rate is 36 kbps over the allocated 25 kHz bandwidth channel (that is 1.44 bits per hertz). Here also digital signal processing is used with vocoders but this time the emphasis is on maximizing data throughput.

Each channel is divided into four time division multiplex (TDM) timeslots in the forward direction. These timeslots are then organized as 'multi- frames' with 18 TDMA frames per multi-frame. For each group of 18 frames, 17 are used for voice and data while the 18th is used for control signaling. This 18th frame is used for the so-called slow associated control channel (SACCA).

In addition to the message content (voice and/or data), the transmitted data includes the protocol information as well as the codes necessary for securing the radio link between subscriber radio and base station. The maximum user data rate per communication channel is 7.2 kbps per timeslot.

For the transmission of voice, the TETRA Coder/Decoder (CODEC) digitizes and compresses the speech in order to minimize bandwidth requirements, and delivers a 4.8 kbps stream of data. This is then carried within the 7.2 kbps channel. The difference of 2.4 kbps is used for protocol overheads, security codes, etc. Data, on the other hand, can be transmitted at different speeds. For extremely high security requirements, data can be transmitted at 2.4 kbps using one time slot. On the other hand, if data protection is disabled, all four time slots can be used to deliver 28.8 kbps.

Two-way conversation is possible, using two corresponding timeslots. Apart from using different timeslots for transmit and receive, the system also uses different frequencies for transmit and receive. The user terminal then continuously switches between transmit and receive frequencies in synchronism with the time allocated slots. The duplex offset (i.e. the difference between the transmit and receive frequencies) are 10 MHz. This means that since transmission and reception does not occur simultaneously, complex filters are avoided.

Because of the large similarity between TETRA and GSM it may seem at first glance that they are competing standards. This is, however, not so. GSM, DECT and TETRA are complementary standards. Whereas GSM is the wireless extension of a digital telephone network, TETRA is the wireless extension of integrated services private branch exchanges. Despite the similarities between GSM and TETRA, the latter covers the needs that GSM cannot fulfil. These include: broadcast calls, group calls, priority calls, open channels, direct mobile to mobile communication, repeater services, and fast call setup (0.3 s).

GSM uses small cells with high capacity and local frequency re-use. TETRA, on the other hand, uses large cells with medium capacity and regional frequency re-use. TETRA call set-up time is 0.3 s, as opposed to several seconds for GSM. Typical calls on TETRA last for 20 s, as opposed to 2 m for GSM systems.

The TETRA system incorporates several gateways to other networks. These include a PSTN gateway for voice calls to and from the public switched telephone network, a PBN gateway (i.e. an X.25 gateway to the public data network) and an ISDN gateway for voice and data into an integrated services digital network.

The defined power classes of TETRA radio equipment are 25 W, 10 W, 3 W and 1 W. The radios can automatically adjust their output power according to the needed field strength.

As this standard has specifically been developed for commercial applications, the transport industry, taxis, couriers, etc. would best be served by this type of radio technology. A number of service utilities (electricity, water, municipalities, etc.) would also be well suited.

11

Wireless LANs

11.1 Introduction

This chapter provides in-depth information about wireless LAN technologies. We begin by addressing the overall concept of wireless networks and the reason for their high rate of deployment worldwide. We then move on to have a look at the various specifications: IEEE 802.11, 802.11b and 802.11a. The other concepts discussed are: OSI layer implementation, medium access control, system components, basic service set and extended service set, ad hoc and infrastructure modes. The security issues concerning wireless networks are also discussed. Finally, the chapter highlights the differences between a WLAN and WPAN (wireless personal area network) and discusses the Bluetooth/IEEE 802.15 standards.

11.1.1 Advantages of wireless data networks

The most important advantage of wireless networking is *mobility*. Wireless data users are free from the restriction of working from a fixed place. They can carry out their work in the conference room, in their project laboratory room, in their car and so on. The only requirement is that the wireless users should be within the specified range of the base station.

The range that is covered by the simplest available equipment is a campus domain. As the equipment gets slightly sophisticated, the range of an 802.11 network can extend up to a few miles.

Wireless networks employ several base stations that facilitate connection to an existing network. The basic requirement in order to provide efficient service is the appropriate functioning of the base stations and antennas. As soon as the infrastructure is in place, it is a very simple task to add a user to the wireless network. The user just needs to be authorized to start utilizing the functionality of the wireless networks.

As in all networks, data transmission in wireless networks takes place over a network medium. The medium in this case is free space, and the transmission method is a type of electromagnetic radiation. In order to be well-suited for deployment on mobile networks, the medium must encompass an extensive area in order to facilitate the movement of the clients throughout that area.

Infrared light and radio waves are the two media that have witnessed the widest use in local area networks. Several portable devices that are available now are equipped with infrared ports. The infrared ports facilitate faster connection to printers and several other peripheral devices.

Infrared light has certain constraints. Walls, partitions and several other constructions can easily obstruct it. Radio waves, on the other hand, find their way through most of the obstructions and thereby facilitate an extensive coverage range.

Thus, a majority of the 802.11 products employ the radio wave physical layer.

11.1.2 Limitations of wireless networking

Wireless networks do not substitute fixed networks. Mobility is advantageous since the network user is accessible even if he/she is on the move. It is essential that servers and other data center equipment must have access to data. However, the physical location of the server is immaterial. Provided that the servers are stationary, they may be connected to wires that are stationary too.

- The wireless network speed is limited by the available bandwidth.
- Wireless network transmission speeds are generally slower when compared to wired systems.
- The security aspect is an important factor that has to be considered for any network. In case of wireless networks, the security concern is enhanced since the network transmissions are accessible to anyone within the transmitter range and with the correct antenna. Additionally, on a wireless network, sniffing is easier because the design of radio transmissions is such that they can be processed by any receiver within the range.

11.2 Specifications IEEE 802.11, 802.11b, 802.11a

The 802.11 working group of the IEEE standards body located in the United States is responsible for stating and maintaining wireless local area network (WLAN) specifications and standardization. The initial IEEE standard 802.11 that was published in 1997 defined three physical (PHY) layer specifications and one medium access control (MAC) specification. Further on, there was work being done in order to extend the initial PHY specifications so that higher data rates could be obtained. This led to the emergence of standards 802.11a and 802.11b. The present published publications are IEEE 802.11-1999, IEEE 802.11a-1999 and IEEE 802.11b-1999.

802.11 networks consist of the following important components. They are:

- *Distribution system*: When numerous access points form a connection that covers a huge domain, communication within themselves is essential in order to trace the movements of mobile stations. The distribution system is the logical component of 802.11 utilized to forward frames to their respective destination. There is no specific technology specified by 802.11 for the distribution system. In a majority of commercial products, the distribution system is executed as a combination of a bridging engine and a distribution medium, which functions as the backbone network utilized to transmit frames between access points. In most of the commercially successful products, Ethernet is utilized as the backbone network technology.
- *Access points*: Frames present on an 802.11 network must be converted to another frame type prior to delivering it to the rest of the world. Devices called access points execute the wireless-to-wired bridging function.
- *Wireless medium*: In order to move frames between stations, the standard utilizes a wireless medium. Numerous physical layers are defined and the

architecture facilitates manifold physical layers to be developed in order to support the 802.11 MAC. Previously, there were just two radio frequency layers and one infrared layer that were standardized.

- *Stations*: Networks transfer data between stations. Stations are computing devices that comprise wireless network interfaces. Generally, stations are battery operated laptop or handheld computers. Let us now get back to the common MAC and the original PHYs prior to discussing the improved data rate versions.

The MAC functions with two network configurations.

1. *Independent configuration*: Communication between stations is direct and there is no infrastructure support.
2. *Infrastructure configuration*: Stations communicate through access points, which belong to a larger distribution system, thereby allowing access to an extended coverage area.

The 802.11 standard specifies three related PHYs. They are:

- Frequency hopping spread spectrum (FHSS)
- Direct sequence spread spectrum (DSSS)
- Baseband IR.

Table 11.1 gives a comparison of the 802.11 standards.

IEEE Standard	Speed	Frequency Band
802.11	1 and 2 Mbps	2.4 GHz
802.11a	Upto 54 Mbps	5 GHz
802.11b	5.5–11 Mbps	2.4 GHz

Table 11.1
Comparison of 802.11 standards

11.2.1   IEEE 802.11b

The new PHY present in IEEE 802.11b is called high-rate direct sequence (HR/DSSS). The 802.11 DDD and 802.11b HR/DSSS PHYs can be present in the same network because the same preamble and header are utilized.

11.2.2   IEEE 802.11a

Nippon telephone and telegraph (NTT) and Lucent technologies proposed the 802.11a PHY. It is based on an orthogonal frequency division multiplexing (OFDM) transmission that operates in the 5 GHz unlicensed national information infrastructure (U-NII) bands. In simple words, OFDM can be stated as a modulation technique that transmits data across several carriers for high data rates at lower symbol rates.

11.3   IEEE 802.11 protocol layer implementation

Protocol layering facilitates research, experimentation and improvements on various parts of a protocol stack.

A vital component of the 802.11 architecture is the physical layer and this is abbreviated as PHY.

11.3.1 Physical layer architecture

The physical layer is divided into two sublayers. They are the physical layer convergence procedure (PLCP) sublayer and the physical medium dependent (PMD) sublayer.

The physical layer logical architecture is illustrated in Figure 11.1.

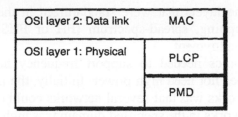

OSI layer 2: Data link	MAC
OSI layer 1: Physical	PLCP
	PMD

Figure 11.1
Physical layer logical architecture

The PLCP acts as the binding agent between the medium access control (MAC) frames and the aerial radio transmissions. It adds its own header. Usually, frames make use of a preamble in order to facilitate synchronization of incoming transmissions. The preamble requirements may depend on the modulation method. However, PLCP adds its own header to any transmitted frames. It is the PMDs responsibility for transmitting the bits it receives from the PLCP into the air employing an antenna.

A clear channel assessment (CCA) function is incorporated by the physical layer to indicate to the MAC when a signal is detected.

11.3.2 The radio link

The initial revision of 802.11 led to the standardization of three physical layers. They are:

- Frequency hopping (FH) spread-spectrum radio PHY
- Direct sequence (DS) spread-spectrum radio PHY
- Infrared light (IR) PHY.

In 1999, two additional physical layers were developed. They are:

1. 802.11a: Orthogonal frequency division multiplexing (OFDM) PHY
2. 802.11b: High rate direct sequence (HR/DS or HR/DSSS) PHY.

11.3.3 IR PHY

802.11 comprises a specification for a physical layer on the basis of infrared (IR) light. Utilizing infrared light instead of radio waves is advantageous. When compared to radio transceivers, IR ports are cheaper. Owing to this, IR ports are standard on every laptop.

IR is highly tolerant to radio frequency (RF) interference because radio waves function at a completely different frequency level. Owing to this, product developers are spared from investigating and complying with directives from the regulatory authorities worldwide.

Security concerns regarding 802.11 are usually concerned with threat of unauthorized users getting connected to the network. Advantages offered by IR band LANs are flexibility and mobility with extremely less security concerns.

However, the range achieved is shorter because IR LANS depend on scattering light off the ceiling.

There has been no product development on the basis of IR PHY. The infrared ports present on laptops have consented to a standards set devised by infrared data association (IrDA) and not 802.11.

11.3.4   802.11 FH PHY

Among all the layers that were standardized in the initial draft of 802.11 in 1997, the frequency hopping, spread-spectrum (FH or FHSS) layer was the first layer to see widespread deployment.

The electronics needed to support frequency hopping modulation is comparatively cheap and does not need high power. Initially, the main advantage of utilizing frequency hopping networks was that several networks could co-exist and the combined throughput of all the networks in the specified domain was high.

Frequency hopping transmission

Frequency hopping transmission is dependent on changing the transmission frequency in a prearranged pattern. This is illustrated in Figure 11.2.

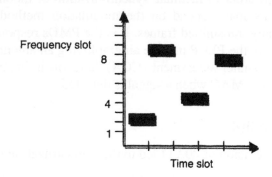

Figure 11.2
Frequency hopping

The vertical axis of the graph partitions the frequency into an array of slots. Similarly, time is also divided into an array of slots. A hopping pattern governs how the slots are utilized. In Figure 11.2, the hopping pattern is {2,8,4,7}. Timing the hops in an accurate manner is the key to success. It is essential that transmitter and receiver must be synchronized in order to ensure that the receiver is always listening on the transmitter frequency.

Frequency hopping is identical to frequency division multiple access (FDMA), but there is a major change from FDMA systems. In FDMA systems, every device is assigned a fixed frequency. Multiple devices share the existing radio spectrum by utilizing various frequencies.

In frequency hopping systems, the frequency is time dependent rather than being fixed. Every frequency is utilized for a small time frame called the 'dwell time'.

11.3.5   802.11 DS PHY

Direct sequence modulation has been the most successful modulation technique utilized with 802.11. The initial 802.11 specification described a physical layer on the basis of a low-speed, direct-sequence spread spectrum. When compared to frequency hopping, direct sequence needs more power in order to obtain the same throughput.

A major advantage of direct sequence transmissions is that the technology can easily adapt itself to higher data rates when compared to frequency hopping network.

Direct sequence transmission is an alternative spread-spectrum technology that can be employed in order to transmit a signal over a wider frequency band. The primary approach of direct sequence technology is to spread the RF energy over a wide band in a continuous way.

Direct sequence modulation functions by applying a chipping sequence to the data stream.

A chip is a binary digit that the spreading process utilizes. Bits are high level data and chips are binary numbers employed in the encoding process. Several chips are utilized in order to encode a single bit into an array of chips.

The bottom line is that direct sequence modulation trades bandwidth for throughput. When compared to conventional narrowband transmission, direct-sequence modulation needs a relatively higher radio spectrum and it is slower. It is easier to obtain high throughput using direct sequence modulation rather than with frequency hopping.

11.3.6   802.11b: HR/DSSS PHY

When the initial 802.11 version was approved in 1997, the major work had just commenced. The initial version of the standard defined FH and DS PHYs, but they could provide data rates only upto 2 Mbps.

2 Mbps is not very useful especially since the transmission capacity has to be equally shared among all the users present in the vicinity. In 1999, the 802.11 Working Group released the second extension to the existing 802.11 specification. This was called 802.11b.

802.11 adds another physical layer into the mix. It utilizes the same MAC as other physical layers and is based on direct-sequence modulation. It permits transmissions upto 11 Mbps, which is suitable for modern networks. Higher data rates resulted in a major commercial success. 802.11 PHY is also called the *high-rate, direct sequence PHY* and is abbreviated as HR/DS or HR/DSSS. Even though the modulation is different, the operating channels are identical to the channels utilized by the low-rate direct sequence.

11.3.7   802.11a: 5 GHz orthogonal frequency division multiplexing (OFDM)

802.11a is based on orthogonal frequency division multiplexing. The OFDM devices utilize one wide frequency channel by splitting it up into numerous subchannels. Every subchannel is utilized for data transmission. Further on, all the slow subchannels are then integrated into one fast subchannel.

The OFDM is similar to the old frequency division multiplexing (FDM). Both the techniques divide the available bandwidth into segments called carriers or subcarriers and ensure that those carriers are always available as definite channels for data transmission. The OFDM increases throughput by utilizing numerous subcarriers in parallel and multiplexing data over the series of subcarriers.

11.4   802.11 Medium access control (MAC)

The 802.11 MAC layer facilitates reliable data delivery for the upper layers over the wireless PHY media. The basis of the data delivery is an asynchronous, best-effort, connectionless delivery of the MAC layer data. Additionally, the 802.11 MAC protects the data that will be delivered by offering security and privacy services.

The primary access method utilized in 802.11 is carrier sense multiple access with collision avoidance (CSMA/CA). The CSMA functions following a 'listen before talk' scheme. This means that a station that wishes to transmit must initially sense the radio channel in order to ascertain if another station is transmitting. If the medium is free, then the transmission may commence.

The CSMA/CA protocol prevents collisions among stations that share the medium by employing some backoff time in case the station's physical or logical sensing medium specifies a busy medium.

In the CSMA/CA scheme, the time gap between frames from a particular user is very minimal. Once a frame has been transmitted from a specific transmitting station, that station must wait until the time gap is up, inorder to try to retransmit. Once the time has elapsed, the station selects a random time (the back-off interval) to wait prior to listening again to check for a clear channel over which transmission has to occur. If the channel is still not free, another back-off interval is selected that is less than the initial back-off time. This process is carried on until the waiting time reaches 0 and the station is allowed to transmit. The above mentioned multiple access process ensures judicious channel sharing thereby avoiding collisions.

Collisions result in the wastage of very precious transmission capacity. Inorder to compensate for this, 802.11 utilizes collision avoidance (CSMA/CA) rather than collision detection (CSMA/CD) utilized by Ethernet.

We will now have an in-depth look at the MAC process with respect to various features.

11.4.1   MAC challenges

Owing to the differences that exist between the wireless network and the wired environment, several challenges have to be faced and the hurdles have to be crossed by the network designers.

RF link quality

On a wired Ethernet, it is perfectly understandable that when a frame is transmitted, we assume that it has been received accurately. Radio links vary, especially so when the frequency utilized is the unlicensed industrial scientific and medical (ISM) band.

Narrowband devices are subject to noise and interference. However, it is essential that unlicensed devices should carry on functioning despite this interference that is bound to exist.

802.11 incorporates positive acknowledgment (ACK). It is essential that every frame must be acknowledged. This is illustrated in Figure 11.3. In case of failure in any part of the transfer, it is considered that the frame is lost.

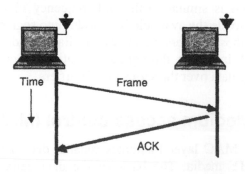

Figure 11.3
Positive acknowledgment of data transmissions

The sequence illustrated in Figure 11.3 is an 'atomic' operation. 802.11 requires stations to restrain contention during atomic operations so that atomic sequences are not disturbed by other stations attempting to utilize the transmission medium.

The hidden node problem

In Ethernet networks, stations rely on reception of transmissions in order to execute the carrier sensing functions of CSMA/CD. Wires present in the physical medium carry signals and distribute them to all network nodes. Wireless networks, on the other hand, have unclear boundaries. In certain situations, the boundaries are so unclear that not all nodes present in the network are in a position to communicate with all other nodes. This aspect is illustrated in Figure 11.4.

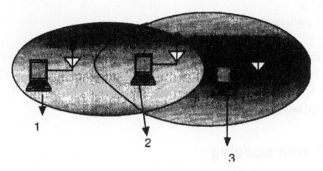

Figure 11.4
Nodes 1 and 3 are hidden

In Figure 11.4, node 2 can communicate with node 1 as well as node 3. The same thing cannot be said of nodes 1 and 3. Node 1 and node 3 cannot communicate directly. From the perspective of node 1, node 3 is a hidden node. If a simple 'transmit-and-pray' protocol was being utilized, it would be very easy for node 1 and node 3 to carry out transmissions at the same time. Additionally, nodes 1 and 3 would not be aware of the error since the collision would be limited to node 2.

Collisions arising from hidden nodes are not easily detected in wireless networks because wireless transceivers are usually half-duplex. A half-duplex node does not transmit and receive simultaneously.

To avoid collisions, 802.11 requires stations to utilize RTS and CTS signals to clean out the area. The procedure is illustrated in Figure 11.5.

In Figure 11.5, node 1 wants to transmit a frame. It commences the process by transmitting a RTS frame. The RTS frame sets aside the radio link for transmission. Additionally, it also quietens any station that hears it. If the target station receives an RTS, it instantly responds back with a CTS. Similar to RTS frame, the CTS frame quietens the stations in close proximity. When the RTS/CTS is complete, node 1 can carry out frame transmission without needing to worry about interference from any hidden nodes. Hidden nodes that are present ahead of the range of the sending station are quietened by the CTS from the receiver. When the RTS/CTS clearing procedure is employed, it is required that any frame received must be acknowledged positively.

The multiframe RTS/CTS transmission procedure takes up a considerable amount of capacity. This is because the additional latency incurred prior to the transmission commencing. Owing to this, it is employed only in high-capacity environments. It is also employed in environments wherein there is a considerable conflict on transmission.

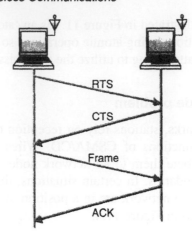

Figure 11.5
RTS/CTS clearing

The RTS/CTS procedure can be controlled by adjusting the RTS threshold if the device driver for the 802.11 card permits the adjustment. The RTS/CTS exchange is carried out for the frames that are larger than the threshold. Frames that are shorter than the threshold are simply transmitted.

11.4.2   Interframe spacing

As in Ethernet, the interframe spacing plays a very vital role in synchronizing access to the transmission medium. 802.11 utilizes four interframe spaces. Three are utilized in ascertaining medium access. Their interrelation is illustrated in Figure 11.6.

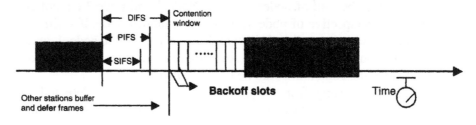

Figure 11.6
Interframe spacing relationships

Different interframe spacings result in the creation of different priorities on the basis of the type of traffic. The logic followed is that high-priority traffic should not wait long after the medium becomes idle. Thus, if there is any high-priority traffic that is waiting, it gets hold of the network prior to low priority frame. In order to aid interoperability between the various data rates, the interframe space is a preset amount of time and it is not based on the transmission speed. However, different physical layers can spell out different interframe space times.

Short interframe space (SIFS)

SIFS is employed for highest priority transmissions like RTS/CTS frames and positive acknowledgments. High-priority transmissions commence once the SIFS has passed by.

Once these high-priority transmissions commence, the medium becomes busy, so the frames that are transmitted after the SIFS has passed by obtain priority over frames that can only be transmitted after longer gaps.

PCF inter frame space (PIFS)

The PIFS is sometimes incorrectly called the priority interframe space. It is utilized by the PCF during conflict-free operation. Stations with data to be transmitted can carry out transmission after the PIFS has gone by and prevent traffic conflicts.

DCF interframe space (DIFS)

The DIFS is the least idle time for conflict-based services. Stations can access the medium immediately in case it is free for a time period greater than the DIFS.

Extended interframe space (EIFS)

There is no mention of EIFS in Figure 11.6 because it is not a constant interval. It is utilized only in case of errors during frame transmission.

Interframe spacing and priority

Atomic operations commence like regular transmissions: they have to wait for the DIFS prior to commencing. Later on, the second and the remaining steps in an atomic operation occur using the SIFS, rather than during the DIFS. This essentially means that second and the remaining steps of an atomic operation seize the medium prior to another frame type being transmitted. By utilizing the SIFS and the network allocation vector (NAV), stations can grab the medium for as long as they want. The network allocation vector is basically a timer that specifies the amount of time for which the medium is reserved.

11.4.3 Frame format

In order to meet the challenges caused by a wireless data link, the MAC was enforced to embrace certain unique features. All the frames do not utilize all the address fields, and the values allocated to the address fields may be transformed on the basis of the MAC frame that is being transmitted.

The basic 802.11 frame is illustrated in Figure 11.7. Fields are transmitted from left to right, and the most significant digit appears last.

We will now have a detailed look at each of the fields present in the 802.11 MAC frame.

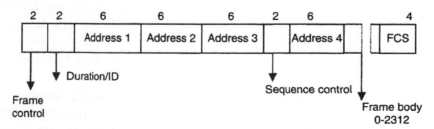

Figure 11.7
Generic 802.11 MAC frame

Frame control

Every frame commences with a two-byte frame-control subfield. The components present in the frame-control subfield are:

- *Protocol version*:   Two bytes denote the version of the 802.11 MAC contained in the rest of the frame. Currently, just one 802.11 version has been developed and has been allocated the protocol number 0.
- *Type and subtype fields*:   The function of the type and subtype fields is the identification of frames utilized. In order to deal with noise and unreliability, several management functions have been integrated into 802.11 MAC. Some of the type and subtype identifiers are indicated below:

Subtype Value	Subtype Name
Management frames (type = 00)[a]	
0000	Association request
0001	Association response
0010	Reassociation request
0011	Reassociation response
0100	Probe request
0101	Probe response
1000	Beacon
1001	Announcement traffic Indian message
1010	Disassociation
1011	Authentication
1100	Deauthentication
Control frames (type = 01)[b]	
1010	Power save poll
1011	RTS
1100	CTS
1101	Acknowledgment (ACK)
1110	Contention-free (CF)-end
1111	CF-END + CF-Ack
Data frames (type = 10)[c]	
0000	Data
0001	Data + CF + Ack
0010	Data + CF-Poll
0011	Data + CF-Ack + CF-poll
0100	Null data (no data transmitted)
0101	CF-Ack (no data transmitted)
0110	CF-Poll (no data transmitted)
0111	Data + CF-Ack + CF-Poll

- *To DS and from DS bits*:   These bits denote whether a frame is intended for the distribution system. All the frames present on infrastructure networks comprise of one of the distribution system's bits set.
- *More fragments bit*:   This bit operates like the 'more fragments' bit in IP. When a high level packet has been fragmented by the MAC, the preliminary

fragment and any subsequent nonfinal fragments set this bit to 1. Certain management frames may be so large that they need fragmentation. The other frames set this bit to 0.

- *Retry bit*: Time and again, frames may be retransmitted. This bit is set to 1 by any of the retransmitted frames in order to help the receiving station eliminate duplicate frames.

- *Power management bit*: Network adapters present on 802.11 are usually built to the PC card to form factor and utilized in battery-powered laptops or handheld computers. In order to increase battery duration, several small devices have the capacity to power down parts of the network interface. This bit specifies whether the sender is in a power saving mode after the present atomic frame exchange has elapsed.

- *More data bit*: In order to assist stations in a power saving mode, access points may buffer frames obtained from the distribution system. An access point which sets this bit to designate at least one frame is present and is addressed to a retired station.

- *WEP bit*: It is a known fact that wireless transmissions can be easily intercepted as compared to fixed network transmissions. 802.11 uses certain encryption routines called wired equivalent policy (WEP) in order to shield and authenticate data. When a frame is processed by WEP, this bit is set to 1 and the frame is altered slightly.

- *Order bit*: The transmission of frames and fragments can be ordered in a particular manner. This will need extra processing by both the sending and receiving MACs.

Duration/ID field

The duration/ID field follows the frame-control field. This field carries out several tasks and assumes one of the three forms as illustrated in Figures 11.8a, 11.8b and 11.8c.

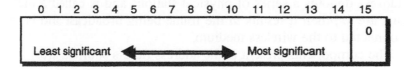

Figure 11.8a
Duration

Figure 11.8b
CFP frames

Figure 11.8c
PS-poll frames

- *Duration setting the NAV*: When bit 15 is 0, the duration/ID field is utilized to set the NAV. The value is denoted in microseconds when the medium is supposed to be busy for the tansmission that is currently under progress. All stations must scrutinize the headers of all the frames that they receive and thereby update the NAV. Any value exceeding the amount of time the medium is busy results in the updating of NAV and blockage of access to the medium for additional time.
- *Frames transmitted during conflict-free periods*: During the conflict-free periods, bit 14 is set to 0 and bit 15 is set to 1. All the other bits are set to 0, So the duration/ID field has a value of 32 768 and this value is interpreted as a NAV.
- *PS-Poll frames*: In PS-Poll frames, bits 14 and 15 are set to 0. Mobile stations may choose to preserve battery power by turning off antennas. It is essential that the sleeping stations must wake up from time to time. In order to ascertain that no frames are lost, stations waking up from their sleep transmit a PS-Poll frame in order to retrieve any buffered frames from the access point. Along with this request, waking stations integrate the association ID (AID) that denotes to the BSS that they represent. The range of the PS-Poll frame is 1–2007. 2008–16 383 are reserved values and are not utilized.

Address fields

An 802.11 frame might comprise up to four address fields. The address fields are numbered since different fields are employed for different purposes on the basis of the frame type. The general rule of thumb is that address 1 is utilized for the receiver, address 2 is utilized for the transmitter, and address 3 is utilized for filtering by the receiver.

Addressing in 802.11 follows the conventions utilized for the other IEEE 802 networks including Ethernet. Addresses are 48 bits in length. If the first bit transmitted to the physical medium is 0, the address denotes a single station. When the first bit is a 1, the address denotes a series of physical stations and is called a multicast address. In case all bits are 1s, then it results in the frame being broadcast and is sent to all stations that are connected to the wireless medium.

The components present in the address field's subfield are:

- *Destination address*: Similar to Ethernet, the destination address is the 48-bit IEEE MAC identifier that corresponds to the final recipient.
- *Source address*: The source address is 48 bits in length and it determines the transmission source. No more than one station can be the source of a frame. Hence, the individual/group bit is always set to 0 to indicate an individual station.
- *Receiver address*: The receiver address is a 48-bit IEEE MAC identifier that denotes the wireless frame. In case it is a wireless station, the receiver address is the destination address. For frames that have to be transmitted to a node on an Ethernet connected to an access point, the receiver is the wireless interface in the access point and the router attached to the Ethernet can be the destination address.
- *Transmitter address*: The transmitter address is 48 bits in length and its function is to determine the wireless interface that transmits the frame onto the wireless medium. The transmitter address is employed only in wireless bridging.

- *Basic Service Set ID (BSSID)*: In order to determine various wireless LANs in the same vicinity, stations may be allocated to a basic service set (BSS). In infrastructure networks, the BSSID is the MAC id that is employed by the wireless interface in the access point. Ad hoc networks generate a random BSSID with the universal/local bit set to 1, in order to avoid conflicts with authorized MAC addresses.

The number of address fields utilized is based on the frame type. A majority of data frames utilize three fields for source, destination and BSSID. The count of the address fields and their respective arrangement in a data frame is based on the frame traversal relative to the distribution system.

Sequence control field

This 16 bit-field is employed for defragmentation and discarding duplicate frames. It comprises a 4-bit fragment number field and a 12-bit sequence number field. Each of the higher level frames are assigned a sequence number as they traverse to the MAC in order to be transmitted. The sequence number subfield functions as a modulo-4096 counter of the transmitted frames. It commences at 0 and is incremented by 1 for every higher-level packet handled by the MAC. In case higher-level packets are fragmented, all packets will contain the same sequence number. When frames are retransmitted, the sequence number remains unaltered. The first fragment is assigned a fragment number of 0. Each following fragment increments the fragment by one. Retransmitted fragments preserve their original sequence numbers in order to aid reassembly.

Frame body

The frame body is also called the data field and it transports the higher-layer payload from station to station. The 802.11 facilitates transmission of frames with a maximum payload of 2 304 bytes of higher-level data.

Frame check sequence

Similar to Ethernet, the 802.11 frame culminates in a frame check sequence (FCS). The FCS is usually referred to as the cyclic redundancy check (CRC) because of the underlying mathematical operations. The FCS enables stations to verify the integrity of the received frames. All fields present in the MAC header and the frame body are incorporated in the FCS. 802.3 and 802.11 employ the similar method in order to calculate the FCS. However, the MAC header employed in 802.11 varies for the header employed in 802.3. Hence, it is essential that the FCS must be recalculated by access points.

11.5   System components

Radio frequency (RF) systems supplement wired networks by extending them. Several different components maybe employed on the basis of frequency and the distance that the signals have to reach. But at the core, all systems typically comprise of several distinct pieces.

Antennas and amplifiers are two RF components that are important for 802.11 users.

Antennas are vital because they are the most substantial feature of an RF system. Amplifiers play the complementing role by facilitating antennas to propel more power, which maybe interesting depending on the nature of 802.11 network that is being built.

11.5.1   Antennas

Antennas are vital for any RF system since they are responsible for the conversion of electrical signals on waves to radio signals and vice versa.

Antennas are represented by a triangular shape as indicated in Figure 11.9.

Figure 11.9
Antenna representations in diagrams

In order to function, the antennas must be made of a conducting material. Radio waves striking an antenna result in the flow of electrons in the conductor and create a current. Similarly, the application of current to an antenna results in an electric field around the antenna. As there is a change in the current to the antenna, the electric field also changes. A changing electric field results in a magnetic field, and the wave is radiated into free space.

The size of the antenna required depends on the frequency. As the frequency increases, the size of the antenna decreases. The shortest antenna that can be devised is ½ wavelength long. An amplitude modulation (AM) station that broadcasts at a frequency of 830 kHz has a wavelength that is above 360 m in length and an equally large antenna.

It is also possible to design antennas depending on the direction preferred.

A large number of antennas can transmit and receive signals from any direction. They are called omnidirectional antennas.

Some applications prefer directional antennas. Directional antennas are those that radiate and receive on a narrower portion of the field.

For a specified input power, a directional antenna can transmit or receive farther with a clearer signal. Directional antennas are very sensitive to radio signals prevailing in the main direction. They are widely used when wireless networks replace wired networks.

802.11 networks utilize omnidirectional antennas for connection at both ends. There might be certain exceptions if the requirement is that network should cover a larger distance. Another important factor to be noted is that in reality there isn't anything that is truly an omnidirectional antenna. We are habituated to think that a vertically mounted antenna is omnidirectional since the signal variation is very minor as we traverse around the antenna in a horizontal plane. But, if we observe the signal that is radiated vertically from the antenna, it is a completely different scenario.

Considering practical antennas for 802.11 devices in the 2.4 GHz band, the wireless PC card contains a built-in antenna.

Most vendors sell an external antenna that plugs into the card. These antennas are nothing extraordinary, and they increase the range of the roaming laptop by a huge margin. There have been several attempts to create high-gain antennas that will aid portable use and can also be used for a base station. There are several commercial antennas available that are specifically designed for employment in 802.11 services.

11.5.2   Amplifiers

Amplifiers amplify signals. Decibels (dB) is the unit used to measure signal boost or gain. Amplifiers can be classified as:

- Low-noise
- High-power
- Everything else.

Low-noise amplifiers (LNAs) are connected to antennas to enhance the received signal such that it can be identified by the electronics of the RF system to which it is connected to. Additionally, they are also evaluated for noise factor. Noise factor can be defined as a measure of external noise introduced to the signal. Smaller noise factors enable the receiver to observe smaller signals, thereby permitting greater range.

High-power amplifiers (HPAs) are utilized to enhance a signal to the maximum degree of power prior to transmission. The unit for measuring output power is dBm (dB relative to 1 mW). Amplifiers radiate heat in addition to amplifying the signal. This is because amplifiers are dependent on thermodynamic laws.

The transmitter present in an 802.11 PC card is low power because it is essential that it runs off a battery if it is installed in a laptop. However, it is possible to install an external amplifier at access points, which can then be connected to the power grid.

802.11 devices are restricted to 1 W of power output and 4 W effective radiated power (ERP). ERP multiplies the transmitter's output power by the antenna gain minus the loss in transmission line. So essentially, if we have a 1-watt amplifier, an antenna that provides 8 dB of gain and 2 dB of transmission line loss, then it results in an effective radiated power (ERP) of 4 W. The total system gain is 6 dB, which multiplies the transmitter's effective power by a factor of 4.

11.6   Basic service set (BSS), extended service set (ESS)

The basic service set (BSS) is the fundamental building block of an 802.11 network. A BSS is a group of stations that communicate with each other. Communications occur in an area called the basic service area.

The basic service area is defined on the basis of the propagation characteristics of the wireless medium. A station present in the basic service area can communicate with other BSS members. BSS can be classified into two categories.

They are:

- Independent BSS
- Infrastructure BSS.

11.6.1   Independent networks

Independent BSS is illustrated in Figure 11.10. Communication takes place directly between IBSS stations. Therefore, it is essential that stations must be within direct communication range.

The smallest possible 802.11 network is an IBSS station that consists of two stations. Usually, an IBSS comprises a few stations that are set up for a particular purpose and for a small time duration. One common use is to create a short-lived network that is capable of supporting a single meeting in a conference room. As the meeting commences, the participants create an IBSS for sharing data. After the meeting concludes, the IBSS is dissolved. Owing to their short duration, small size and focused purpose, IBSSs are referred to as ad hoc BSSs or ad hoc networks.

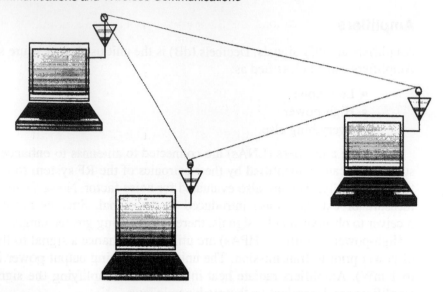

Figure 11.10
Independent BSS

11.6.2   Infrastructure networks

An infrastructure BSS is illustrated in Figure 11.11.

Infrastructure networks are distinguished from independent networks by the utilization of an access point. Access points are employed for communication in Infrastructure networks, including communication between mobile nodes in the same service area.

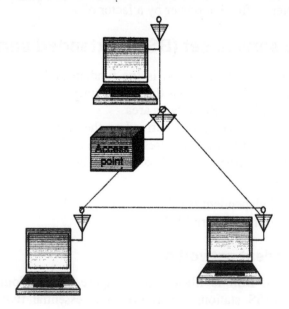

Figure 11.11
Infrastructure BSS

If a mobile station present in an infrastructure BSS wishes to communicate with a second mobile station, two hops must be undertaken by the communicator.

During the first hop, the frame is transmitted to the access point by the originating mobile station. During the second hop, the frame is transmitted to the destination station by the access point.

Since all communications are transferred via an access point, the basic service area relevant to an infrastructure BSS is defined by the points in which transmission from the access point can be received.

Multihop transmission requires more transmission capacity when compared to a directed frame.

However it has the following advantages:

- The basis of definition for an infrastructure BSS is the distance from the access point. However, there is no restriction placed between mobile stations themselves. Permitting direct communication between mobile stations would result in saving the capacity, but at the expense of augmented physical layer complexity since it would be essential that mobile stations preserve relationships with other mobile stations present in the service area.
- Access points in infrastructure networks have the capacity to aid stations that are looking forth to save power. Access points can note when a station enters a power saving mode and buffer frames for the same. Battery operated stations have the capacity of turning the wireless transceiver off and power it up only to transmit and retrieve buffered frames from the access point.

In an infrastructure network, stations must associate with an access point in order to get new services. Association can be defined as the process by which a mobile station joins an 802.11 network. It corresponds to plugging in the network cable on an Ethernet. It is not a symmetric process. The association process is always triggered by the mobile station. The 802.11 standard places no restriction on the number of mobile stations that an access point may cater for. Implementation considerations may restrict the number of mobile stations that an access point may serve. However, in reality, the low throughput of wireless networks is likely to restrict the number of stations placed on a wireless network.

11.6.3   Extended service areas

BSSs can facilitate coverage in small organizations and homes. However, they fail to provide network coverage when the coverage area becomes too large. 802.11 facilitates the creation of wireless networks of a large size by linking BSSs into an extended service set (ESS).

An ESS comprises several BSSs chained together with a backbone network. 802.11 does not spell out a specific backbone technology. All it requires is that the backbone should offer the required services.

An ESS is an amalgamation of the four BSSs. It is essential that the access points should be configured such that they belong to the same ESS.

In real world operation, the overlap degree between the BSSs would be much higher. It is essential that continuous coverage is offered within the extended service area. Secondly, it is not desirable that users walk past the area enveloped by BSS3 when passing from BSS1 to BSS2.

Communication is possible between stations present within the same ESS, even if they belong to dissimilar basic areas and move about between basic service areas. If stations within an ESS have to communicate, the wireless medium must act like a single layer connection. Access points function as bridges, so direct communication between an ESS calls for the network layer to be a two layer connection. Any link layer connection is adequate to meet the requirements.

The numerous access points in a single area might be connected to a single hub or switch, or they can utilize virtual LAN if the link layer connection must cover a large area.

Access points (APs) in an ESS function to facilitate the outside world to utilize a single MAC address to communicate with a station located within the ESS.

11.7   IEEE 802.11 architecture

There are two different ways by which a network can be configured as per the specifications in IEEE standard for wireless LANs.

They are:

1. Ad hoc
2. Infrastructure.

In case of an ad hoc network, computers form a network on the spur of the moment. Refer to Figure 11.12. It is clearly visible that the network does not have a fixed structure. There are no fixed points existing and every node can communicate with every other node.

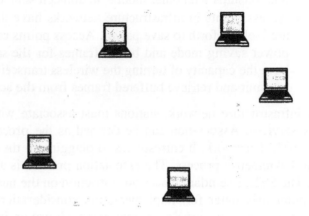

Figure 11.12
Ad hoc network

The second type of network structure used in wireless LANs is the infrastructure network. It is indicated in Figure 11.13 This architecture utilizes fixed network access points with which mobile nodes can communicate. Sometimes, these network access points are connected to landlines inorder to increase the LAN's capability. This is accomplished by wireless nodes to the wired nodes. This structure is rather identical to the current cellular networks that exist worldwide.

Figure 11.13
Infrastructure network

11.8   IP roaming

The 802.11 performs some very intelligent jugglery with MAC addresses. Stations communicate with a MAC address as if it is stationary, similar to an Ethernet station. Rather than being put up in a fixed location, however, access points keep track of the mobile station when it is in close vicinity and transmit frames from the wired network to it over airwaves.

It is immaterial which access point the mobile station correlates with primarily because the relay function is executed only by the suitable access point. The station on the wired network communicates with the mobile station as if it were attached to the wire directly.

Mobile IP executes a similar trick with IP addresses. The external world utilizes a single IP address that is stationary in a location called the home location. Instead of being serviced by a user's system, the IP address present at the home location is serviced by the home agent. Identical to the access point, it is the function of the home agent to maintain information about the present location of the mobile node.

When the mobile node is 'at home', packets can be easily sent to it. In case the mobile node gets to a different network (also called the foreign network or visited network), the foreign location has to be registered with the home agent. The home agent is thus able to redirect the traffic from the home address to the mobile node located on the foreign network.

Consider an example, wherein two wireless LANs are developed on different IP subnets. On its home subnet, it is possible for a wireless station to transmit and receive traffic easily because it is on its home network.

When the wireless station traverses from its home subnet to the second subnet, it attaches itself to the network employing the usual procedure. It gets connected to an access point and requests for an IP address using dynamic host configuration protocol (DHCP). On a wireless station wherein Mobile IP is not employed, connections are disrupted at this juncture owing to the sudden change in IP address invalidating the state of all TCP connections.

Wireless stations using Mobile IP software can maintain their connection state by registering with the home agent. The mobile station can receive packets for the mobile station, verify its registration tables, and then transmit packets to the mobile station at its present location. The mobile station comprises of two addresses. One address is the home address that it can utilize for connections that were set up using home networks. Additionally, it might also utilize the address allocated on the foreign network. No TCP state is ever invalidated because the mobile station never ceased to utilize its home address.

In the above-mentioned approach, several security problems are apparent. The most important among them being the authentication of protocol operations and the security of the redirected packets while traversing from the home network to the mobile station's present location.

The task of preserving precise routing information, the traditional forwarding tables at the Internet gateways and Mobile IP agents is a vital challenge. Finally, the protocol must function with both IPv4 and IPv6.

11.9   Security issues

The word 'broadcast' assumes a completely new meaning in wireless networks. Security concerns have been a major reason for concern about 802.11 deployments ever since the standardization effort commenced. In order to address spying concerns, the IEEE emerged with the wired equivalent privacy (WEP) standard, which is present in clause 8.2

of 802.11. WEP can be utilized by stations in order to protect data as it passes thorough the wireless medium. There is no protection for data once it passes through the access point.

The WEP was introduced basically to provide a security solution for wireless LANs. However, the WEP design has several drawbacks and hence it cannot be utilized at all in numerous cases. The WEP encryption code was broken in late 2001. The reason for this was some concealed problem with the cryptographer cipher utilized by WEP.

Cryptographer cipher is a new term for some readers and might sound very confusing. We will now look at WEP's cryptographic heritage and its design flaws.

11.9.1   WEP cryptographic operations

Communications security has three important aims. It is extremely essential that any protocol that claims to provide secure data as it traverses across a network must assist network managers to obtain the following goals.

They are:

1. Confidentiality
2. Integrity
3. Authentication

Confidentiality denotes the data that is protected from interruption by malicious parties.
Integrity denotes the data that is unmodified.
Authentication bolsters the security strategy because a portion of data reliability is based on its origin.

Users must ascertain that that data is transmitted from the actual source from which it was claimed be transmitted. Systems must employ authentication in order to safeguard data in the precise manner.

WEP offers certain operations that aid the above stated aims. Confidentiality is supported by frame body encryption. While data is traversing, it is protected by an integrity check that facilitates receivers to check for the correctness of data when it is received.

Owing to defects in the RC4 cipher, confidentiality cannot be obtained. The design of the integrity check is extremely poor. Authentication is carried out on the MAC addresses of the users and not the users themselves.

Additionally, frames are encrypted by WEP as they traverse in a wireless medium. There is no protection offered to frames as they traverse on a wired backbone. WEP has been designed to protect the data from external trespassers. However, once the WEP key is identified, the wireless medium is comparable to a big shared network.

11.9.2   WEP data processing

Prior to encryption, the frame has to undergo an integrity check algorithm. The integrity check algorithm results in the generation of a hash called an *Integrity Check Value* (ICV). The function of the ICV is to protect the contents against tampering by ascertaining that the frame hasn't been modified while traversing.

WEP clearly states that a 40 bit secret key must be utilized. The secret WEP key is merged with a 24-bit *Initialization Vector* (IV) in order to create a 64-bit RC4 key

The first 24 bits of the RC4 key constitute the initialization vector (IV), followed by the 40-bit WEP key. RC4 takes the 64 input bits and produces a key stream that is equivalent in length to the frame body and the IV. An XOR operation is then carried out on the frame body and the IV inorder to cipher it.

11.9.3 WEP keying

In order to offer protection to the traffic from malicious decryption attacks, WEP employs a series of up to four default keys. Additionally, it might also utilize pairwise keys called mapping keys, if permitted. Default keys are distributed to all the stations that exist in a service set. Once a station gets the default keys for its service set, communication is possible using WEP.

One of the weaknesses of cryptographic protocols is key reuse. Owing to this reason, WEP comprises of an alternative class of keys that is employed for pairwise communications. These keys are shared only between two communicating stations. The two stations sharing a key have a key mapping relationship.

11.9.4 Problems associated with WEP

There are certain flaws that the cryptographers have identified in WEP. The designers stated the employment of RC4, which is acknowledged as a strong cryptographic cipher. Attackers do not restrict themselves to just attacking the cryptographic algorithms. They also tend to attack the weaknesses in the cryptographic system.

Design flaws

WEPs design flaws came to light when the Internet security, applications, authentication and cryptography (ISSAC) group at the University of California, Berkeley, published the initial results on the basis of their study on the WEP standard.

Some of the problems that they encountered are stated below:

- WEP utilizes a CRC (cyclic redundancy check) for the integrity check. Even though the integrity check value is encrypted by the RC4 keystream, CRCs are not secure cryptographically. Employing a weak integrity check does not avert attackers from modifying frames.
- It is a fact that standardized WEP provides a shared secret of only 40 bits. Many vendors like to claim otherwise. Security experts have frequently stated that just 40 bits is in no position to protect sensitive data. They have also stated that at least 128-bit keys are essential to provide at least some form of security.
- Stream ciphers are susceptible to analysis whenever the keystream is reutilized. WEP's employment of the IV provides a hint to an attacker regarding the reuse of a keystream. Two frames sharing the identical IV will definitely use the same secret key and keystream. This problem is aggravated implementations that might or might not pick random IVs.
- Infrequent re-keying facilitates attackers to form decryption dictionaries. A large series of encrypted frames are along with the same keystreams. As several frames with the same IV are gathered, additional information can be obtained about the encrypted frames even though the secret key is not recovered. Owing to the overburdened typical system and network administration, infrequent re-keying is an established rule.

11.9.5 Conclusions and recommendations

The design of WEP was basically to offer a fair degree of protection to frames in the air. It was never ever designed for scenarios that call for high security level. The IEEE 802.11

working group has set up a group that completely concentrates on the security aspect. The task group is looking into an enhanced security standard. In the meantime, several vendors are providing corrective approaches that facilitate stronger public-key authentication and random session keys. However, these are all temporary solutions and will not prove to be beneficial in the long run. It is possible to develop better solutions from off-the-shelf standardized components.

Stated below are certain conclusions and recommendations:

- Managing the key manually is a vital problem. Peer-to-peer networking systems have all the associated problems in the area of management and scalability. WEP too is no different. Employing pairwise keys is a major burden on system administration and does not provide much security in any way whatsoever.
- It is a known fact, that a secret is no longer a secret when it is shared widely. WEP follows the principle of sharing a secret key openly. Users come and go, and WEP keys must be changed with every departure, in order to be certain that protection is provided by WEP.
- Confidential data must employ effective cryptographic systems that have been extensively designed keeping in mind the security aspect. The most preferred choices are IPSec or SSH. The reason for preference of a certain choice is technical evaluation, product availability, user expertise and many non-technical factors like pricing, licensing and institutional acceptance.
- WEP does not offer protection to users from one another. When all the users have the WEP key, any kind of traffic can be decrypted easily. Wireless networks that offer protection to users from each other must employ applications that have a very strong security back up.

11.10 WLANs (wireless local area networks) vs WPANs (wireless personal area networks)

The key differences between IEEE specifications for WLAN (IEEE 802.11) and WPAN (IEEE 802.15) are:

WLAN	WPAN
Environment	
It is appropriate for applications that require MAC provisions (ability to support large numbers of nodes involved in concurrent communication, mechanism to avoid interference) and high range (10–100+ meter) necessary to support LAN applications	WPANs are appropriate for applications which are limited to the WPAN range (0–10 m), do not require the complexity of a LAN-style MAC scheme, and are extremely sensitive to power consumption and per-unit cost

Range

WLANs support segments at least 100 m in length. The specifications for WPANs focus on short-distance wireless networks with a range of 0–10 m.

Medium Access Control Mechanism (MAC)	
IEEE 802.11 calls for a carrier sense multiple access with collision avoidance (CSMA/CA) MAC scheme	The Task Group 1 (TG1 - Bluetooth) specification calls for a time division multiple access (TDMA) scheme. This mechanism provides no 'listen before transmission' type of interference avoidance as in CSMA/CA
Cost Factor	
No specific cost constraints for 802.11	Cost factors are critically important in the WPAN arena; for example, the unit cost to manufacturers for TG1 (Bluetooth) devices is targeted in the $10 (or under) range in volume
Power Consumption	
No specific power consumption constraints for 802.11	802.15 standards are intended to address power consumption issues critical to this class of small portable/mobile device

11.11   Wireless PANs: Bluetooth/IEEE 802.15

The 802.15 wireless personal area network (WPAN) group was established in the summer of 1999 and focuses on developing standards for short distance wireless networks utilized for networking portable and mobile computing devices like PCs, cell phones, speakers, microphones and several other consumer electronics. The main purpose of 802.15 was to create an IEEE based PAN standard that would complement the work of 802.11.

The term 'PAN' defines new development in WLANs, where the vital factors are low implementation costs, low power consumption and ease of utilization.

The 802.15 task group had an agreement with the Bluetooth SIG and it was decided that the 802.15 task group would adopt the Bluetooth v1.0b standard as the basis for its work. The initial task assigned to 802.15 task group was to rework on the standard and ensure that it fits into the IEEE networking model. This initial task had to be completed by the end of 2000.

In January 2000, the IEEE set up another task group, called 802.15.2. The main task assigned to this group was to conduct a thorough study regarding the co-existence and interoperability between 802.15 WPANs and 802.11 WLANs. After carrying out a thorough study, it is required that the group should suggest ways that would facilitate coexistence. Additionally, the group must also suggest alterations to other 802.11 or 802.15 standards in order to augment co-existence with other wireless devices that operate in unlicensed frequency bands.

In March 2000, the third task group was set up. It was called 802.15.3 and the main purpose behind its creation was to aid in the creation of a high-rate WPAN standard. Companies like Cisco, Motorola and Eastman Kodak suggested that this group should be created. This group is responsible for the creation of physical (PHY) and medium access control (MAC) specifications in order to facilitate high data rate wireless connectivity between a stationary, moveable, and moving devices that are present within a personal operating space (POS) or the devices entering a personal operating space. POS refers to the space relevant to a person or object that ranges up to 10 m in all directions and encompasses the person/object whether in motion or stationary.

The ultimate aim of the standard is to obtain an interoperability level with 802.15.1 and 802.15.2 and other wireless devices. The data transmission rate is high: 20 Mbps or more in order to comply and satisfy the consumer requirements of the consumer multimedia industry for wireless PAN communications. Additionally, 802.15.3 will also look into the quality of service (QoS) capabilities that are essential to support multimedia data types like video applications and digital imaging.

11.12   Protocol stack

One of the unique features of Bluetooth is that it offers a complete protocol stack that facilitates various applications to communicate over a wide array of devices. The protocol stack for voice, data, and control signaling is illustrated in Figure 11.14.

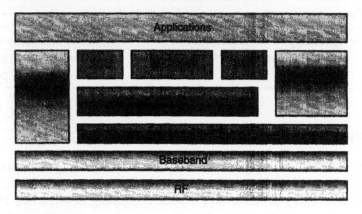

Figure 11.14
Protocol stack for Bluetooth

The RF layer specifies the radio modem utilized for transmission and reception of information. The baseband layer specifies the link control both at the bit and the packet level and it also states coding and encryption for frequency hopping and packet assembly operation. The link management protocol (LMP) arranges the links to their devices by offering authentication and encryption, traffic scheduling and packet format. The logical link control and adaptation protocol (L2CAP) offers connection-oriented as well as connectionless data services to the upper layer protocols. These services encompass segmentation and reassembly, protocol multiplexing and group abstractions for data packets up to 64 kB in length. The audio signal is transmitted directly from the application to the baseband.

Different applications might employ different protocol stacks but all of them share the similar physical and data link control mechanisms. Above the L2CAP, there are three other protocols.

The service discovery protocol (SDP) discovers the characteristics of the services and connects two or more Bluetooth devices in order to support services like printing, faxing, e-commerce facilities and teleconferencing.

The telephony control protocol specification provides telephony services.

The RFCOMM provides an RS-232-like serial interface. Numerous non-Bluetooth specific protocols can be implemented on the Bluetooth devices to support legacy applications by utilizing RFCOMM.

11.12.1   The OSI reference model

The open systems interconnect (OSI) reference model is illustrated in Figure 11.15. Bluetooth does not exactly resemble the model. However, we will try to relate to the different parts of the Bluetooth stack to the different parts of the OSI model.

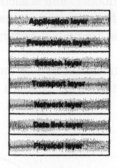
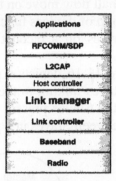

Figure 11.15
OSI reference model and Bluetooth

The Physical layer is in charge for the electrical interface to the communications media, including modulation and channel coding. It therefore covers the radio part of the baseband.

The data link layer is in charge of transmission, framing and error control over a specific link.

The network layer is in charge of data transfer across the network, unrelated to the media and the precise topology of the network. This comprises the higher end of the link controller, setting and maintaining multiple links, and also looks after the link manger tasks.

The transport layer is in charge of reliability and multiplexing of data transfer across the network to the level that is offered by the application. Therefore, it overlaps at the high end of the LM and covers the host controller interface (HCI). The HCI facilitates actual data transfer mechanisms.

The session layer is responsible for providing management and data flow control services that are covered by the L2CAP and the lower RFCOMM/SDP ends.

The presentation layer offers a common representation for application layer data by supplementing service structure to the data units, which is the main task of RFCOMM/SDP.

The application layer manages communications between host applications.

11.13   Piconets and scatternets

Prior to actually discussing the piconet and scatternet topologies, we need to understand the concepts of master and slave.

If devices are to hop to new frequencies after very packet, it is essential that they must all agree on the frequency series that they will utilize. Bluetooth devices can function in two modes, that is, either as a master or as a slave. The master establishes the frequency hopping sequence. Slaves follow the master and synchronize correspondingly to the master in time and frequency.

Every Bluetooth device comprises a distinct Bluetooth device address and a Bluetooth clock. When slaves get connected to the master, they are informed about the Bluetooth device address and the clock of the master. They then utilize this information in order to calculate the frequency hop sequence.

Apart from controlling the frequency hop sequence, the master decides as to when devices should transmit. The Master permits Slaves to transmit by providing slots for voice or data traffic.

We shall now move on to the terms piconets and scatternets.

A group of devices functioning together with one common Master is called as a piconet. All devices on a piconet abide by the frequency hopping sequence and timing of the Master.

Figure 11.16 depicts a point-to-point connection on the left and a piconet comprising three slaves talking to the master, on the right. This is a point-to-multipoint connection, where the Slaves present in a piconet are only linked to the Master. There are no direct links between Slaves in a piconet.

Figure 11.16
Point-to-point piconet and point-to-multipoint piconet

The Slaves that can be present in a piconet is restricted to seven as per the specification, with every Slave communicating with the shared Master. However, if a larger area has to be covered, several piconets can be linked together to form a scatternet.

If a device is present in more than one piconet, then it has to essentially spend some slots in one piconet and some slots in the other piconet. Figure 11.17 is a scatternet where one device is a slave in two piconets.

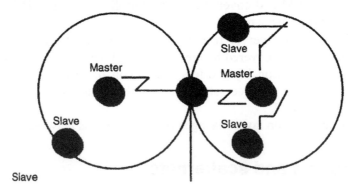

Figure 11.17
Scatternet

11.14   Medium access control

The modulation technique and the operating frequency of the Bluetooth radio system is quite similar to frequency hopping spread spectrum (FHSS) used in 802.11.

However, the MAC mechanism in Bluetooth differs greatly from that used in 802.11. The Bluetooth access mechanism is a voice-oriented pioneering system that does not have any similarity either with carrier sense multiple access with collision avoidance (CSMA/CA) or the voice-oriented code division multiple access (CDMA) and time division multiple access (TDMA) access methods, but still it comprises certain elements that have a resemble to these access methods.

The medium access mechanism of Bluetooth is a fast frequency hopping-code division multiple access/time division duplex (FH-CDMA/TDD) system that utilizes polling in order to establish the link. The fast hopping of 1600 hops per second facilitates slots of 625 μs (625 bits at 1 Mbps) 'for one' packet transmission that permits fast performance in the presence of interference. Bluetooth is a CDMA system that is implemented employing FHSS.

In the Bluetooth CDMA, every piconet has its own spreading sequence. However, in the DSSS/CDMA systems utilized for digital cellular systems, every user link can be recognized with a spreading code. The reason for the non-selection of DSSS/CDMA for Bluetooth is because DSSS/CDMA requires central power control that is virtually impossible in the scattered ad hoc topology visualized for Bluetooth applications.

Without any requirement for centralized power control for CDMA operation, the FH/CDMA in Bluetooth facilitates overlapping of several piconets in the same area, thereby offering an effective throughput that is much larger than 1 Mbps.

In Bluetooth, the FH/CDMA approach is preferred over simple FDM or OFDM because ISM bands at 2.4 GHz permit only spread spectrum technology. The access point in each piconet is TDMA/TDD. The TDMA format permits participation of multiple voice and data terminals in a piconet. The TDD abolishes cross-talk existing between transmitter and receiver thereby facilitating a single chip implementation in which the radio oscillates between transmitter and receiver modes.

In order to share the medium across a larger number of terminals, at every slot 'M' decides and polls a 'S'. Polling is employed over contention access methods because contention results in a large amount of overhead for the short packets (625 bits) that were chosen in order to implement a fast FHSS system.

11.15   Frame formats

The basis of the Bluetooth packet format is one packet per hop and a basic 1-slot packet of 625 μs that can be stretched to three slots (1875 μs) and five slots (3125 μs). This frame format and the FH/TDMA/TDD access mechanism facilitate an 'M' terminal to poll several 'S' terminals at diverse data rates for voice and data applications in order to form a piconet.

The entire Bluetooth packet structure is illustrated in Figure 11.18. The access bit field comprises 74 bits; the header field comprises 54 bits. There are about 2744 bits for various payloads that can be as long as five slots. Bluetooth utilizes highly adaptable shorter packets for easy incorporation and enhanced performance in fading. However, these are gains that are obtained at the cost of an increased overhead percentage that minimizes throughput.

As illustrated in Figure 11.18, the access code field comprises a four-bit preamble, a four-bit trailer and a 64-bit synchronization PN-sequence with numerous codes comprising good autocorrelation and cross-correlation properties. PN is the abbreviation for port negotiation. It is the RFCOMM command employed for configuring a data link connection. RFCOMM is the protocol for RS-232 serial cable imitation.

The 48-bit IEEE MAC address that is distinctive to every Bluetooth device is utilized as the starting point to obtain the PN-sequence for hopping frequencies of the device.

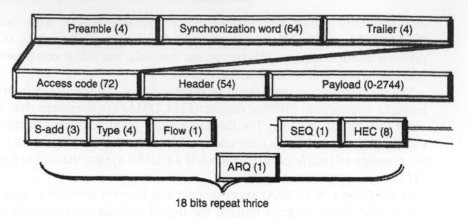

Figure 11.18
Entire frame format of Bluetooth packets

Access codes can be categorized into four types:

1. The first type recognizes a 'M' terminal and its associated piconet address.
2. The second type states an 'S' identity that is utilized to page a specific 'S'.
3. The third type is a fixed access code meant primarily for the inquiry process. We shall look into the details about inquiry later on in this chapter.
4. The fourth type is the dedicated access code that is set aside for identifying a particular set of devices like fax machines, cellular phones or printers.

As illustrated in Figure 11.18, the header field comprises 18 bits that are represented thrice with a 1/3 forward error correction (FEC) code. FEC is an error correction employed to offer data protection on some Bluetooth packets. The 18 bits are arranged in the following manner.

The initial 3 bits represent an 'S' addressing followed by 4-bit packet type. Further on, there is the 3-bit status reports, and an 8-bit error check parity.

The 3-bit S-ADD permits addressing the seven probable active 'M's in a piconet. The 4-bit packet type permits 16 options for various grade-voice services, data services at dissimilar rates, and four control packets. The 3-bit status reports are utilized in order to indicate overflow of the terminal with information, acknowledgment of successful packet transmission, and sequencing in order to distinguish between the sent and resent packets.

The Bluetooth Special Interest Group (SIG) states the different payloads and associated packet type codes that facilitate implementation of several voice and data services. Different master–slave pairs present in a piconet can utilize different packet types and the packet type may be altered randomly during the course of a communication session.

16 different packet formats have been identified by the four-bit packet type for the payloads of the Bluetooth packets. Six of these payloads are asynchronous connectionless (ACL), utilized mainly for packet data communications. Three of the packet formats are synchronous connection oriented (SCO), utilized mainly for voice communications. One is an integrated voice and data packet. Four are control packets that are common for SCO as well as ACL links.

The three SCO packets illustrated in Figure 11.19 are high-quality voice (HV) packets numbered as HV1, 2 and 3 in order to indicate the quality level of the service. The SCO packets are single slot packets. The length of the payload is constant at 240 bits, and they do not utilize the status report bits, but they are conveyed over reserved periodic duplex intervals in order to support 64 kbps per voice user. HV1 utilizes all 240 bits for user voice samples. HV2 utilizes 160 bits for user voice samples and 80 bits of parity for a 1/3 FEC code, and HV3 utilizes 80 bits of user voice samples and 160 bits of parity for a 2/3 FEC code. In order to maintain the data rate for voice samples at 64 kbps, the HV1, HV2, and HV3 packets in each direction are transmitted every six, four and two slots.

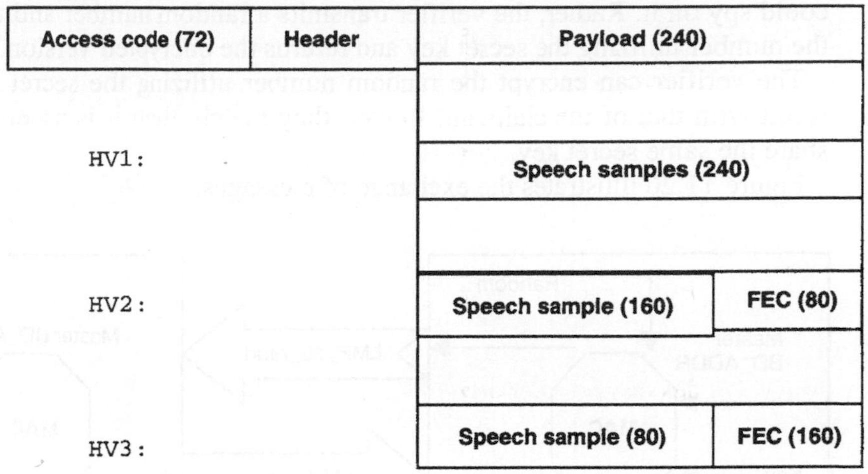

Access code (72)	Header	Payload (240)

| HV1: | Speech samples (240) | |

| HV2: | Speech sample (160) | FEC (80) |

| HV3: | Speech sample (80) | FEC (160) |

Figure 11.19
SCO-1 slot packet frame formats

11.16   Security issues

Cable based communications is extremely secure when compared to wireless communications.

The aspect of security is dealt with at many levels in the Bluetooth specifications:

- The baseband specification elaborates the SAFER+ algorithms utilized for security procedures.
- The link manager specification deals with the coverage of link level procedures for configuring security.
- The host control interface (HCI) spells out in detail about how security related events are reported by a Bluetooth module to its host.
- The generic access profile looks after the coverage of security modes and user-level procedures for utilization in all products implementing Bluetooth profiles.
- The Bluetooth SIG has issued a white paper on the security architecture that suggests a framework for security implementation and provides examples of how services might utilize security.

The Bluetooth specification utilizes a variant of the secure ad fast encryption routine (SAFER+) cipher in order to authenticate devices.

The first requirement for authentication is that the encryption engine must be initialized to a random number. After initialization, the encryption engine requires four inputs. They are:

- A number for encryption or decryption (this is the data that is being transmitted between devices)
- The Master's Bluetooth device address
- The Master's Bluetooth slot clock
- A secret key that is shared between the two devices.

All devices in a piconet are aware of the Master's Bluetooth device address and slot clock. The secret key utilized for encryption differs from time to time. Sometimes, a device needs to confirm that it actually shares a secret key with another device that is claiming to share the key.

It is not possible for the verifier to ask the claimant to transmit the key since anybody could spy on it. Rather, the verifier transmits a random number and the claimant encrypts the number utilizing the secret key and returns the encrypted version.

The verifier can encrypt the random number utilizing the secret key and compare its result with that of the claimant. In case they match, then it is ascertained that both sides share the same secret key.

Figure 11.20 illustrates the exchange of messages.

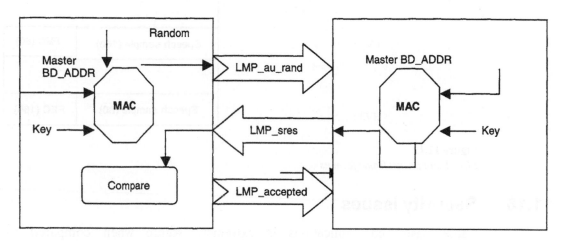

Figure 11.20
Authentication utilizing the Bluetooth encryption engine

11.17   Using Bluetooth

Bluetooth is different from a wired network because there is no need to attach a cable to the devices you are communicating with. Sometimes, you might be unaware as to which device you are communicating with. In order to cope with this, Bluetooth provides inquiry and paging mechanisms and a service discovery protocol (SDP).

11.17.1   Discovering Bluetooth devices

Consider two Bluetooth enabled devices like a cellular phone and a laptop computer. The cellular phone is capable of functioning as a modem utilizing the dial up networking profile, and it scans periodically to check if anyone wants to utilize it.

The laptop user opens an application that requires a Bluetooth dial-up networking connection. To utilize this application, the laptop is aware that it has to establish a Bluetooth link to a device that supports the dial-up networking profile. The first step in establishing the connection is to find out about the Bluetooth enabled devices in the nearby vicinity. Thus, the laptop executes an inquiry to look out for Bluetooth enabled devices in the neighborhood. To accomplish this, the laptop transmits a sequence of enquiry packets. Finally, the cellular phone answers with a frequency hop synchronization (FHS) packet.

The FHS packet comprises of the information required by the laptop to get connected to the cellular phone. Additionally, it contains the device class of the cellular phone, that is, the major and minor parts.

The major device class informs the laptop that it has located a phone. The minor device class provides information that the phone located is a cellular phone.

The sequence of events occurring next is completely based on the application design.

11.17.2 Connecting to a service discovery database

To ascertain if a device supports a specific service, the application will have to connect to the device and utilize the service directory protocol (SDP).

Initially, the laptop pages the cellular phone utilizing the information that it has collected during inquiry. In case the phone is scanning for pages, it responds back and an asynchronous connectionless (ACL) baseband can be established to transfer data between the two devices.

Once the ACL connection has been set up, a logical link control and adaptation protocol (L2CAP) connection can be established across it. An L2CAP connection is utilized whenever data has to be transmitted between Bluetooth devices. L2CAP facilitates many protocols and services to utilize one baseband ACL link. L2Cap differentiates between various protocols and services utilizing an ACL connection by adding a protocol and service multiplexer (PSM) to every L2CAP packet. The PSM varies for every protocol. Since the connection will be utilized for service discovery, its protocol/service multiplexer (PSM) = 0x0001, a special value that is always utilized for the service directory.

The laptop utilizes the L2CAP channel to set up a connection to the service discovery server on the cellular phone. The laptop's service discovery client can then query the cellular phone's service directory server in order to transmit all the information relevant to the dial-up networking profile. The service directory present on the cellular phone looks throughout the database and returns the attributes corresponding to dial-up networking.

After the service directory information has been obtained, the laptop might suddenly decide to close down the connection to the cellular phone. In case the laptop wishes to gather service directory information from several devices present in the area, it seems sensible to close down all the links, since one device can utilize only a limited number of links at a given point of time and a lot of battery power will be consumed if the links are kept alive.

Once the laptop has accumulated service directory information from devices in the area, the next step is again up to the application's discretion.

11.17.3 Connecting to a Bluetooth service

The paging process that sets up a baseband ACL link is similar to the one that was utilized when connecting for service discovery. At this juncture, the link is being established for a protocol that might call for specific service requirements. Therefore, the application functioning on the laptop might wish to configure the link in order to meet its requirements. This is accomplished by the application conveying its requirements to the Bluetooth module utilizing the host controller interface. Further on, the module's link manager configures the link utilizing the link management protocol.

Once the ACL connection has been established and is up to the laptop's liking, an L2CAP connection is established. The dial-up networking employs RFCOMM, an RS-232 emulation layer, so the L2CAP connection utilizes the protocol stack multiplexer for RFCOMM (PSM = 0x0003).

Once the L2CAP link has been established, an RFCOMM connection can be set up across it. Every protocol or service is provided its own channel number. The cellular phone's channel number for dial-up networking was transmitted to the laptop in the service discovery information. Therefore, the laptop has an idea as to which channel number it should employ when setting up the RFCOMM connection.

Finally, the dial-up networking connection is established utilizing the RFCOMM connection, and the laptop can commence to utilize the dial up networking services of the cellular phone. It is now possible for the laptop to utilize the cellular phone in order to establish connections across the phone network with the need for a data cable completely eliminated.

Appendix A

Practical session data

A.1   Fiber-optic design exercise

Design a fiber-optic system to link two sites 4 km apart.
Cable is supplied in 1 km lengths.

Data

Fiber data

MM GI	850 nm	1310 nm
Fiber loss	3.5 dB/km	1.0 dB/km
Transmitter launch Power into fiber 62.5/125	−10 dBm	−15 dB
Receiver sensitivity	−27 dBm	−35 dB
Receiver dynamic range	15 dB	15 dB
Connector loss	0.5 dB	0.5 dB (per connector end)
Splice loss	0.2 dB	0.2 dB (per splice)

Tasks

1. Sketch system schematic.
2. Identify how many splices and connectors will be needed.
3. Calculate link power budgets for 850 and 1310 nm.
4. Which system do you use?

A.2   Microwave design exercise

A. For the microwave path defined by the instructor, use the Teledes program to calculate the required antenna heights at both sites. Use the transmit frequency of 2.49 GHz for these calculations.
B. For the above microwave radio path, select the appropriate components from the following data sheets for the design of a system to achieve 99.9% availability.

Assumes

TX1	2.495 GHz
TX2	2.485 GHz
Both transmitters	0.5 Watt output
Hybrid loss	3.0 dB
Receiver sensitivity	−85 dBm
Total connector loss	1.0 dB for each Tx, Rx chain at each site.

Task

1. Sketch equipment schematic for both sites.
2. Select an appropriate 'Grid Pak' antenna for each site.
3. Select and appropriate feeder type for each site, either coax or waveguide.
4. Calculate the fade margin using the Teledes program.

A.3   Network design exercise

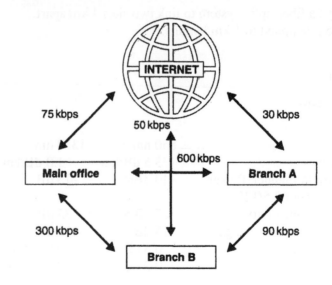

Figure A.1
Traffic between three branches and the Internet

Assumptions

1. Minimum hop routing is to be used.
2. Required integral number of 64 kbps and 2.048 Mbps line.
3. Design on basis that link usage is not to exceed 50% of available link bandwidth.

Data cost

Terminal router	$US 2000
Transit router	$US 3500
CSU/DSU	$US 500
Monthly 64 kbps rental	$US 200
Monthly 2.048 Mbps rental	$US 1250

Task

1. Sketch least cost system configuration.
2. Determine what leased lines are needed.
3. Show termination equipment and traffic carried on each route.

A.4 'Putting it all together'

Conceptual design of communications infrastructure for company XYZ Pty Ltd

Company XYZ has a factory in Johannesburg and regional sales offices in Pretoria, Cape Town and Durban.

The sale offices are located in one building each, with floor areas measuring approximately 30 m × 30 m each. They initially require 10 telephone lines each, as well as 10 PCs with Internet access. These figures may eventually double. The PCs need to be networked locally, and it has been decided to use Internet 100BaseTX for this purpose. To minimize costs, voice over IP telephone communications has to be used wherever possible to make and receive calls on all communications within the company. It must be possible to make and receive calls on all PCs. If VoIP services are unavailable, regular PSTN services are to be used for communication with the outside world.

The factory comprises five buildings spread over a 3 km × 3 km area. Each building requires 100 telephone extensions and 50 networked PCs with Internet access. Once again, 100BaseTX is to be used. The same VoIP requirements apply. Managers (approximately 20 in total) must be able to make and receive telephone calls anywhere on site.

Managers also need high-speed access (>1 Mbps) to the corporate network from home with their laptops, keeping security considerations in mind.

All corporate networked devices must be incorporated in one virtual private network.

Tasks

1. Sketch a rough schematic for the network/telephone solution for company XYZ. Make assumptions wherever necessary, but state them.
2. Make recommendations for the appropriate access technology to be used for the managers' houses, as well as for the intra-corporate WAN technology (i.e. the 'cloud') to be used.
3. Where possible, 'size' devices such as PBXs, key systems, hubs, etc. Where catalogs are not available, ask your instructor for guidance.

Have fun!

A.5   Fiber-optic design Solution (A.1)

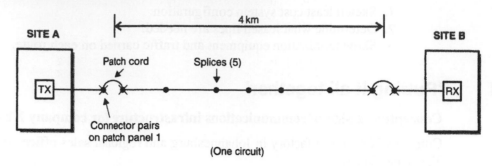

Figure A.2
Schematic

1. Assume five splices are required. Three are needed because the cable comes in 1 km lengths, plus two more because you may not be able to install the 1 km length without cutting the cable.

   Assume eight connectors are in each circuit. Standard practice is to install cable on patch panel at each end, to provide flexible access to all fibers in cable. Each of the four connections to the patch cords involves two connectors.

2.

Link power budget calculations	850 nm	1310 nm
Launch power (Tx)	−10 dBm	−15 dBm
Receiver sensitivity (RSL)	−27 dBm	−35 dBm
System gain (power budget)	17 dB	20 dB
Fiber loss (4 km)	14 dB	4 dB
Connector losses (8)	4 dB	4 dB
Splice losses (5)	1 dB	1 dB
TOTAL losses	19 dB	9 dB
Excess power over system gain	−2 dB	11 dB
Required link margin	6 dB	6 dB
Acceptable system	NO	YES

3. Dynamic range calculation @ 1310 nm

   Maximum allowable receiver power = RSL + dynamic range

   $$= -35\,dBm + 15\,dB$$

   $$= 20\,dBm$$

   Maximum power at receiver $= -15\,dBm - 9\,dB = -24\,dBm$

   There is therefore no problem with dynamic range.

   On a shorter cable with fewer connectors, it may be necessary to install attenuators to avoid dynamic range problems.

A.6   Microwave design Solution (A.2)

1. Minimum size 'grid-pak' antenna required.
   KP4-24 – verify in Teledes.
2. Coaxial cable is acceptable.
   LDF SP-50A-2 – verify in Teledes.

3. See solution on software disk. 'Example MLD'
   Copy into Teledes directory and view using Teledes 'Microwave Link Design'
   sheet, using File 'Load' command.

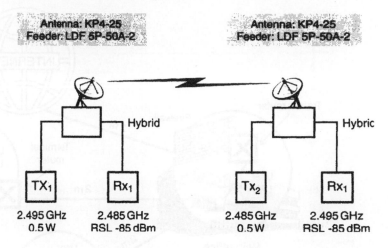

Figure A.3
Equipment layout

A.7   Network design Solution (A.3)

Aggregate traffic (Figure A.4) requires minimal routes and needs a maximum of one router hop to reach all destinations. Consolidation of Internet traffic enables effective use of corporate firewall at the main office.

Using requirement that link usage not exceed 50% of available bandwidth allows us to arrive at the circuit quantities shown.

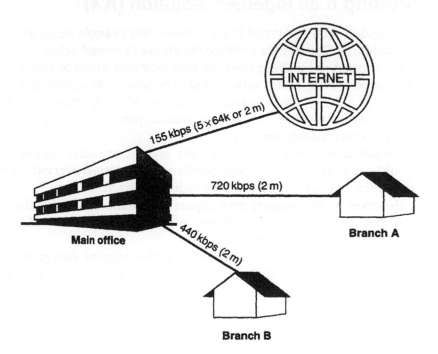

Figure A.4
Aggregate traffic

Note: Branch B is best served by 2 m data stream rather than 14, 64 k circuits.

Similarly on costings given it is probably more cost effective to provide 2 m to Internet, rather than 5 × 64 k (Figure A.5).

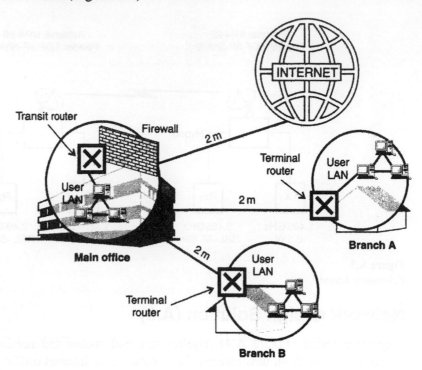

Figure A.5
Equipment configuration

A.8   'Putting it all together' Solution (A.4)

In common with many real life situations, this example needs greater definition to arrive at a solution. Some of the required details are discussed below.

Floor plans of buildings showing user locations would be required to effectively run the structured cabling. Sales offices should be able to be served by a single hub using Cut 5 cable. A factory poses greater problems. Multiple hubs/switches would be required depending on size and traffic requirements, particularly if Ethernet LAN is connected to process control equipment.

Requirements for managers to make and recieve calls anywhere on site is vague and would need clarification. Local mobile communication everywhere within a 3 km × 3 km site is very difficult, unless cellular radio is utilized.

However most managers may require communication within their own factory building and not necessarily all over the site. In this case one of the cordless telephone technologies may be suitable.

High-speed access for managers at home requires further details, particularly cable distance. The use of DSL technologies is indicated here, but speed is dependant on distance.

Appendix B
Glossary

A

3GPP	Third generation partnership project for mobile telephony.
10Base2	IEEE 802.3 (or Ethernet) implementation on thin coaxial cable (RG58/Au).
10Base5	IEEE 802.3 (or Ethernet) implementation on thick coaxial cable.
10BaseT	IEEE 802.3 (or Ethernet) implementation on unshielded 22AWG twisted-pair cable.
A1-Net	Austrian name for GSM 900 networks.
Access control mechanism	The way in which the access to the physical transmission medium is managed by the LAN.
ACD	Automatic call distribution.
Address	A normally unique designator for location of data or the identity of a peripheral device, which allows each device on a single communications line to respond to its own message.
ARP	Address resolution protocol. A TCP/IP process used by a router or a source host to translate the IP address into the physical hardware address, for delivery of the message to a destination on the same physical network.
ADSL	Asymmetric digital subscriber line.
AIOD	Automatic identification of outward dialed calls.
Alias frequency	A false lower frequency component that appears in data reconstructed from original data acquired at an insufficient sampling rate (less than twice the maximum frequency of the original data).
AM	Amplitude modulation. A modulation technique whereby the amplitude of a high frequency sinusoidal carrier is modulated by the signal (e.g. audio) to be transmitted.
ASK	Amplitude shift keying. A variation of AM used for transmitting data across an analog network, such as a switched telephone network. The amplitude of a single (carrier) frequency is varied or modulated between two levels, one for binary 0 and one for binary 1.
AMPS	Advanced mobile phone system. A first generation mobile phone system primarily used in the USA. Uses frequency division multiple access (FDMA).
Analog	A continuous real time phenomena where the information values are represented in a variable and continuous waveform.

ANSI	American National Standards Institute. The national standards development body in the USA.
AP	Access point for wireless LANs (in effect a wireless hub connected to the system backbone).
API	Application programing interface.
Application layer	The highest layer of the seven layer ISO/OSI reference model structure, which acts as an interface between user applications and the lower layers of the stack.
ARP	Address resolution protocol.
ARPANET	The packet switching network, funded by the DARPA, that has evolved into the worldwide Internet.
ARP cache	A table of recent mappings of IP addresses to the physical addresses, maintained in each host and router.
AS	(1) Australian Standard. (2) Autonomous system.
ASCII	American Standard Code for Information Interchange. A universal standard for encoding alphanumeric characters into 7 or 8 binary bits.
ASIC	Applications-specific integrated circuit.
ASN.1	Abstract syntax notation one. An abstract syntax used to define the structure of the protocol data units associated with a particular protocol entity.
Asynchronous	Communications where characters can be transmitted at an arbitrary, unsynchronized point in time and where the time intervals between transmitted characters may be of varying lengths. Communication is controlled by start and stop bits at the beginning and end of each character.
ATM	Asynchronous transfer mode. A fast cellular switching system which uses 53-byte cells for transmission of data over wide and local area networks.
Attenuation	The decrease in the magnitude of strength (or power) of a signal. In cables, generally expressed in dB per unit length.
Attenuator	A passive network that decreases the amplitude of a signal (without introducing any undesirable characteristics to the signals such as distortion).
AUI cable	Attachment unit interface cable for 10Base5. Sometimes called the drop cable. Attaches hosts to transceiver unit (MAU).
AWG	American wire gage.

B

Balanced circuit	A circuit so arranged that the impressed voltages on each conductor of the pair are equal in magnitude but opposite in polarity with respect to ground.
Bandwidth	The range of frequencies available expressed as the difference between the highest and lowest frequencies is expressed in Hertz (or cycles per second). Also used as an indication of capacity of the communications link.
Base address	A memory address that serves as the reference point. All other points are located by offsetting in relation to the base address.
Baseband	Baseband operation is the direct transmission of a signal (e.g. voice or data) data over a transmission medium without the prior modulation on a high frequency carrier band.

Baud	Unit of signaling speed derived from the number of events per second (normally bits per second). However if each event has more than one bit associated with it the baud rate and bits per second are not equal.
BCC	Block check character. Error checking scheme with one check character; a good example being block sum check.
BCD	Binary Coded decimal. A code used for representing decimal digits in a binary code.
B-CDMA	Broadband CDMA, now known as W-CDMA.
BERT/BLERT	Bit error rate/block error rate testing. An error checking technique that compares a received data pattern with a known transmitted data pattern to determine transmission line quality.
BGP-4	Border gateway patrol-4. An exterior gateway protocol, developed by Cisco, and currently the de facto standard for routing between autonomous systems.
BIOS	Basic input/output system.
Bipolar	A signal range that includes both positive and negative values.
BIT (BINARY DIGIT)	Derived from 'BInary DigiT', a one or zero condition in the binary system.
Bits per sec (BPS)	Unit of data transmission rate.
Block sum check	This is used for the detection of errors when data is being transmitted. It comprises a set of binary digits (bits) which are the modulo 2 sum of the individual characters or octets in a frame (block) or message.
Bluetooth	Short distance wireless access technology (2.4 GHz band).
BNC	Bayonet nut connector. Bayonet type coaxial cable connector.
Bridge	A device to connect similar sub-networks without its own network address. A store-and-forward device used mostly to segment networks.
Broadband	Opposite of baseband. In broadband operation the data to be transmitted is first modulated on a high frequency carrier signal. It can then be simultaneously transmitted with other data modulated on different carrier signals on the same transmission medium.
Broadcast	A message on a bus intended for all devices which requires no reply.
BS	British Standard.
	Basic service set (IEEE 802.11 wireless LANs).
BSC	Bisynchronous transmission. A byte or character-oriented communication protocol that has become the industry standard (created by IBM). It uses a defined set of control characters for synchronized transmission of BCD between stations in a data communications system.
Buffer	An intermediate temporary storage device used to compensate for a difference in data rate and data flow between two device (also called a spooler for interfacing a computer and a printer).
Burst mode	A high speed data transfer in which the address of the data is sent followed by back to back data words while a physical signal is asserted.
Bus	A data path shared by many devices with one or more conductors for transmitting signals, data or power.
Byte	A term referring to eight associated bits of information; sometimes called a 'character' or 'octet'.

C

Capacitance (mutual)	The capacitance between two conductors with all other conductors, including shield, short-circuited to the ground.
Capacitance	Storage of electrically separated charges between two plates having different potentials. The value is proportional to the surface area of the plates and inversely proportional to the distance between them.
Cascade	Two or more electrical circuits in which the output of one is fed into the input of the next one.
CCITT (see ITU-T)	Consultative Committee International Telegraph and Telephone. An international organization that sets worldwide telecommunications standards (e.g. V.21, V.22, V.22bis).
CDMA	Code division multiple access (IS-95). A cellular technology based on spread-spectrum techniques.
cdma2000	New second-generation CDMA specification.
CDPD	Cellular digital packet data.
CENTREX	Central Office exchange server.
Character	Letter, numeral, punctuation, control figure or any other symbol contained in a message.
Characteristic impedance	The impedance that, when connected to the output terminals of a transmission line of any length, makes the line appear infinitely long. The ratio of voltage to current at every point along a transmission line on which there are no standing waves.
CLOCK	The source(s) of timing signals for sequencing electronic events e.g. synchronous data transfer.
CMRR	Common mode rejection ratio.
CMV	Common mode voltage.
CNR	Carrier to noise ratio. An indication of the quality of the modulated signal.
CODEC	Coder and decoder. Used to convert analog speech into digital signal and vice versa.
Collision	The situation when two or more LAN nodes attempt to transmit at the same time.
Common mode signal	The common voltage to the two parts of a differential signal applied to a balanced circuit.
Common carrier	A private data communications utility company that furnishes communications services to the general public.
Conditioned lines	Leased circuits with equalization to improve data transmission performance.
Contention	The facility provided by the dial network or a data PABX which allows multiple terminals to compete on a first come, first served basis for a smaller number of computer posts. Also used as a medium access method by Ethernet/IEEE 802.3 networks where multiple network interface cards contend for access to the bus.
GPRS	General packet radio service.
CRC	Cyclic redundancy check. An error checking mechanism using a polynomial algorithm based on the content of a message frame at the transmitter and included in a field appended to the frame. At the receiver, it is then compared with the result of the calculation that is performed by the receiver.

Cross-talk	A situation where a signal from a communications channel interferes with an associated channel's signals.
CSMA/CD	Carrier sense multiple access/collision detection. When two stations transmit at the same time on a LAN, they both cease transmission and signal that a collision has occurred. Each then tries again after waiting for a pre-determined time period. This contention-based mechanism forms the basis of the Ethernet/IEEE 802.3 specifications.
CSC	Circuit Switched cellular.
CSD	Circuit Switched data.
CT-2	Second-generation digital cordless standard.
CT-3	Third generation cordless standard, similar to DECT.
CTI	Computer telephony integration.
CTS	Cordless telephone system for GSM phone systems.

D

D-AMPS	Digital AMPS (IS-54), aka North American TDMA.
DASK	Differential amplitude shift keying.
Data link layer	This corresponds to layer 2 of the ISO reference model for open systems interconnection. It is concerned with the reliable transfer of data (no residual transmission errors) across the data link being used.
Datagram	A type of service offered on a packet-switched data network. A datagram is a self-contained packet of information that is sent through the network with minimum protocol overheads.
DCS-1800	A different version of GSM, operating in the 1800 MHz band.
DDS	Digital data service. Leased digital circuits supplied by telecom service provider.
Decibel (dB)	A logarithmic measure of the ratio of two signal levels where $dB = 20\log 10V1/V2$ or where $dB = 10\log 10P1/P2$ and where V refers to voltage or p refers to power. Note that it has no units of measure.
Decoder	A device that converts a combination of signals into a single signal representing that combination.
DECT	Digital European Cordless Telephone Standard.
Default	A value or setup condition assigned, which is automatically assumed for the system unless otherwise explicitly specified.
Delay distortion	Distortion of a signal caused by the frequency components making up the signal having different propagation velocities across a transmission medium.
DES	Data encryption standard.
DFSK	Differential phase shift keying.
DID	Direct inward dialing.
Dielectric constant (E)	The ratio of the capacitance using the material in question as the dielectric, to the capacitance resulting when the material is replaced by air.
DIGITAL	A signal which has definite states (normally two).
DIN	Deutsches Institut Fur Normierung. German National standards agency.
DIP	Acronym for dual-in-line package referring to integrated circuits and switches.

DMA	Direct memory access. A technique of transferring data between the computer memory and a device on the computer bus without the intervention of the microprocessor. Also abbreviated to DMA.
DOD	Direct outward dialing.
DPSK	Differential phase shift keying.
Driver software	A program that acts as the interface between a higher level coding structure and the lower level hardware/firmware component of a computer.
DTMF	Dual tone multi-frequency. Audio signaling technique used with touch-tone telephones.
Duplex	The ability to send and receive data simultaneously over the same communications line.
DWDM	Dense wavelength division multiplexing.
Dynamic range	The difference in decibels between the overload or maximum and minimum discernible signal level in a system.

E

E-1	Digital circuit operating at 2,048 Mbps. Corresponds to 30-channels in the European digital hierarchy.
EBCDIC	Extended binary Coded decimal interchange Code. An eight-bit character code used primarily in IBM equipment. The code allows for 256 different bit patterns.
EDAC	Error detection and correction.
EEPROM	Electrically erasable programmable read only memory. Non-volatile memory in which individual locations can be erased and re-programed.
EIA	Electronic Industries Association. A standards organization in the USA specializing in the electrical and functional characteristics of interface equipment.
EIA-232-C	Interface between DTE and DCE, employing serial binary data exchange. Typical maximum specifications are 15 m at 19 200 baud.
EIA-422	Interface between DTE and DCE employing the electrical characteristics of balanced voltage interface circuits.
EIA-423	Interface between DTE and DCE employing the electrical characteristics of unbalanced voltage digital interface circuits.
EIA-449	General-purpose 37-pin and 9-pin interface for DCE and DTE employing serial binary interchange.
EIA-485	The recommended standard of the EIA that specifies the electrical characteristics of drivers and receivers for use in balanced digital multipoint systems.
EISA	Enhanced industry standard architecture.
EMI/RFI	Electromagnetic interference/radio frequency interference. 'Background noise' that could modify or destroy data transmission.
EMS	Expanded memory specification.
	The activation of a function of a device by a defined signal.
Encoder	A circuit which changes a given signal into a coded combination for purposes of optimum transmission of the signal.
EPROM	Erasable programmable read only memory. Non-volatile semiconductor memory that is erasable in ultra violet light and reprogrammable.

Equalizer	The device that compensates for the unequal gain characteristic of the signal received.
Error rate	The ratio of the average number of bits that will be corrupted to the total number of bits that are transmited for a data link or system.
ESS	Extended service set (IEEE 802.3 Wireless LANs).
Etherloop	Ethernet over the local loop. High speed customer access technology.
Ethernet	Name of a widely used LAN, based on the CSMA/CD medium access method (IEEE 802.3).
ETSI	European Telecommunications Standardization Institute.

F

Farad	Unit of capacitance whereby a charge of 1 C produces a 1 V potential difference.
FCC	Federal Communications Commission.
FCS	Frame check sequence. A general term given to the additional bits appended to a transmitted frame or message by the source to enable the receiver to detect possible transmission errors.
FDDI	Fiber distributed data interface.
FDM	Frequency division multiplexing.
FIFO	First in, first out.
Filled cable	A cable construction in which the cable core is filled with a material that will prevent moisture from entering or passing along the cable.
FIP	Factory instrumentation protocol.
Firmware	A computer program or software stored permanently in PROM or ROM or semi-permanently in EPROM.
Flame retardancy	The ability of a material not to propagate flame once the flame source is removed.
Floating	An electrical circuit that is above the earth potential.
Flow control	The procedure for regulating the flow of data between two device preventing the loss of data once a device's buffer has reached its capacity.
FM	Frequency modulation.
Frame	The unit of information transferred across a data link. Typically, there are control frames for link management and information frames for the transfer of message data.
Frequency	Refers to the number of cycles per second.
FSK	Frequency shift keying.
FTTC	Fiber to the curb.
FTTH	Fiber to the home.
Full-duplex	Simultaneous two-way independent transmission in both directions. See Duplex.

G

G	Giga (metric system prefix–109).
Gateway	A device to connect two different networks which translates the different protocols.
G.Lite	Most common ADSL standard based on ITU-T G.992.1.

GMSS	Geostationary Mobile Satellite Standard, a satellite air interface developed from GSM.
Ground	An electrically neutral circuit having the same potential as the earth. A reference point for an electrical system also intended for safety purposes.
GSM	Global system for mobile communications, a.k.a. Groupe Speciale Mobile. Mobile system that operates in the 900 or 1800 MHz band.

H

H.323	The UTU-T standard for packet-based multimedia communication systems (VoIP).
Half-duplex	Transmissions in either direction, but not simultaneously.
Hamming distance	A measure of the effectiveness of error checking. The higher the Hamming distance (HD) index, the safer is the data transmission.
Handshaking	Exchange of pre-determined signals between two devices establishing a connection.
HDLC	High level data link control. The international standard communication protocol defined by ISO to control the exchange of data across either a point-to-point data link or a multi-drop data link.
HDR	High data rate a.k.a. $1 \times$ EV. A wireless Internet access technology based on CDMA.
HDS	High-speed digital subscriber line.
Hertz (Hz)	A term replacing cycles per second as a unit of frequency.
Hex	Hexadecimal. Numbering system to base 16. The 4-bit binary numbers are represented by digits 0–9, A–F.
HFC	Hybrid fiber coax.
HLR	Home location register.
Host	This is normally a computer belonging to a user that contains (hosts) the communication hardware and software necessary to connect the computer to a data communications network.
HUMAN	High speed unlicensed metropolitan network.

I

I/O address	A method that allows the CPU to distinguish between different hardware components in a system.
ICMP	Internet control message protocol.
IEC	International electrotechnical commission.
IEE	Institution of electrical engineers.
IEEE	Institute of electrical and electronic engineers. An American-based international professional society that issues its own standards and is a member of ANSI and ISO.
IFC	International FieldBus Consortium.
Impedance	The total opposition that a circuit offers to the flow of alternating current or any other varying current at a particular frequency. It is a combination of resistance R and reactance X, measured in ohms.

Inductance	The property of a circuit or circuit element that opposes a change in current flow, thus causing current changes to lag behind voltage changes. It is measured in henrys.
Insulation resistance (IR)	That resistance offered by an insulation to an impressed DC voltage, tending to produce a leakage current though the insulation.
Interface	A shared boundary defined by common physical interconnection characteristics, signal characteristics and measurement of interchanged signals.
Interrupt handler	The section of the program that performs the necessary operation to service an interrupt when it occurs.
Interrupt	An external event indicating that the CPU should suspend its current task to service a designated activity.
IP	Internet protocol.
IR	Infrared.
ISA	Industry standard architecture (for IBM personal computers).
ISDN	Integrated services digital network. Telecommunications network that utilizes digital techniques for both transmission and switching. It supports both voice and data communications.
ISO	International Standardization Organization.
ITU	International Telecommunications Union.
IVR	Integrated voice recognition.

J

Jabber	Garbage that is transmitted when a LAN node fails and then continuously transmits.
JTAPI	Java telephony application programing interface.
Jumper	(1) A wire connecting one or more pins on one end of a cable only. (2) A connection between two pins on a circuit board to select an operating function.

K

k (kilo)	This is 210 or 1024 in computer terminology, e.g. 1 kb = 1024 bytes.

L

LAN	Local area network. A data communications system confined to a limited geographic area typically about 3 km with high data rates (4 to 155 Mbps).
LCD	Liquid crystal display. A low power display system used on many laptops and other digital equipment.
Leased (or private) line	A private telephone line without interexchange switching arrangements.
LED	Light emitting diode. A semiconductor light source that emits visible light or infrared radiation.
Line driver	A signal converter that conditions a signal to ensure reliable transmission over an extended distance.

Linearity	A relationship where the output is directly proportional to the input.
Link layer	Layer 2 of the ISO/OSI reference model. Also known as the data link layer.
LMDS	Local multipoint distribution system.
LLC	Logical link control (IEEE 802.2).
Loop resistance	The measured resistance of two conductors forming a circuit.
Loopback	Type of diagnostic test in which the transmitted signal is returned on the sending device after passing through all, or a portion of, a data communication link or network. A loop-back test permits the comparison of a returned signal with the transmitted signal.

M

m	Meter. Metric system unit for length.
M	Mega. Metric system prefix for 106.
MAC	Media access control (IEEE 802).
MAN	Metropolitan area network.
Manchester encoding	Digital technique (specified for the IEEE-802.3 Ethernet baseband network standard) in which each bit period is divided into two complementary halves; a negative to positive voltage transition in the middle of the bit period designates a binary '1', whilst a positive to negative transition represents a '0'. The encoding technique also allows the receiving device to recover the transmitted clock from the incoming data stream (self-clocking).
Mark	This is equivalent to a binary 1.
MAU	(1) Media access unit (Ethernet 10Base5). (2) Multistation access unit (IBM token ring).
MCU	H.323 Multipoint control unit.
MDF	Main distribution frame.
MFC	Multifrequency compelled. Interexchange signaling system using two out of five audible tones.
Microwave	AC signals having frequencies of 1 GHz or more.
MMDS	Multipoint microwave distribution service.
MMS	Manufacturing message services. A protocol entity forming part of the application layer. It is intended for use specifically in the manufacturing or process control industry. It enables a supervisory computer to control the operation of a distributed community of computer-based devices.
MNP	Microcom networking protocol. An error correction and data compression protocol used by modems.
MODEM	MODulator – DEModulator. A device used to convert serial digital data from a transmitting terminal to a signal suitable for transmission over a telephone channel or to reconvert the transmitted signal to serial digital data for the receiving terminal.
MOS	Metal oxide semiconductor.
MOV	Metal oxide varistor.
MSU	Mobile subscriber unit e.g. a mobile (cell) phone.
MSC	Mobile switching center.

MTBF	Mean time between failures.
MTSO	Mobile telephone switching office.
MTTR	Mean time to repair.
Multidrop	A single communication line or bus used to connect three or more points.
Multiplexer (MUX)	A device used for division of a communication link into two or more channels either by using frequency division or time division.

N

NAMPS	Narrowband AMPS. An enhancement of the AMPS system to increase its call capacity.
Narrowband	Typically frequencies below 1.5 Mbps.
Network architecture	A set of design principles including the organization of functions and the description of data formats and procedures used as the basis for the design and implementation of a network (ISO).
Network layer	Layer 3 in the ISO/OSI reference model, the logical network entity that services the transport layer responsible for ensuring that data passed to it from the transport layer is routed and delivered throughout the network.
Network topology	The physical and logical relationship of nodes in a network; the schematic arrangement of the links and nodes of a network typically in the form of a star, ring, tree or bus topology.
Network	An interconnected group of nodes or stations.
Node	A point of interconnection to a network.
Noise	A term given to the extraneous electrical signals that may be generated or picked up in a transmission line. If the noise signal is large compared with the data carrying signal, the latter may be corrupted resulting in transmission errors.
Non-linearity	A type of error in which the output from a device does not relate to the input in a linear manner.
NRZ	Non-return to zero. Pulses in alternating directions for successive 1 bits but no change from existing signal voltage for 0 bits.
NRZI	Non-return to zero inverted.

O

OC-1	Optical carrier level 1. 51.84 Mbps. The lowest optical rate in the SONET standard.
OFDMA	Orthogonal frequency division multiple access.
OHM (W)	Unit of resistance such that a constant current of 1 A produces a potential difference of 1 V across a conductor.
Optical isolation	Two networks with no electrical continuity in their connection because an opto-electronic transmitter and receiver has been used.
OSI	Open systems interconnection.
OSPF	Open shortest past first. An interior gateway (i.e. routing) protocol.

P

Packet	A group of bits (including data and call control signals) transmitted as a whole on a packet switching network. Usually smaller than a transmission block.
PAD	Packet assembler/disassembler. An interface between a terminal or computer and a packet switching network.
Parallel transmission	The transmission method where data is sent simultaneously over separate parallel lines. Usually unidirectional such as the Centronics interface for a printer.
PBX	Private branch exchange.
PCF	Point control function. A polling type of medium access control used by IEEE 802.11 wireless LANs for real-time data.
PCM	Pulse code modulation. The sampling of a signal and encoding the amplitude of each sample into a series of uniform pulses.
PCMCIA	Personal Computer Memory Card Industries Association. Standard interface for peripherals for laptop computers.
PCS	Personal communication service (IS-136) based on D-AMPS (IS-54).
PDH	Plesiochronous digital hierarchy.
PDU	Protocol data unit.
Peripherals	The input/output and data storage devices attached to a computer e.g. disk drives, printers, keyboards, display, communication boards, etc.
Physical layer	Layer one of the ISO/OSI reference model, concerned with the electrical and mechanical specifications of the network termination equipment.
PLC	(1) Programmable logic controller, (2) Power line carrier.
PLL	Phase locked loop.
Point-to-point	A connection between only two items of equipment.
Polyethylene	A family of insulators derived from the polymerization of ethylene gas and characterized by outstanding electrical properties, including high IR, low dielectric constant, and low dielectric loss across the frequency spectrum.
PVC	Polyvinyl chloride. A general purpose family of insulation materials whose basic constituent is polyvinyl chloride or its copolymer with vinyl acetate. Plasticizers, stabilizers, pigments and fillers are added to improve mechanical and/or electrical properties of this material.
PORT	(1) A place of access to a device or network, used for input/output of digital and analog signals. (2) A number used by the transmission control protocol to identify individual processes (programs) running on a computer.
Presentation layer	Layer 6 of the ISO/OSI reference model, concerned with negotiation of a suitable transfer syntax for use during an application, for example the translation of EBCDIC to ASCII. Encryption is also handled at this level.
Protocol	A formal set of conventions governing the formatting, control procedures and relative timing of message exchange between two communicating systems.
PM	Phase modulation.
PSDN	Public switched data network. Any switching data communications system, such as Telex and public telephone networks, which provides circuit switching to many customers.

PSTN	Public switched telephone network. This is the term used to describe the (analog) public telephone network.
PSK	Phase shift keying a.k.a. Binary phase shift keying (BPSK).
PTT	Post, telephone and telecommunications authority.
PVC	(1) Poly vinyl chloride. (2) Permanent virtual circuit.

Q

QAM	Quadrature amplitude modulation.
QoS	Quality of service.

R

R/W	Read/write.
RAM	Random access memory. Semiconductor read/write volatile memory. Data is lost if the power is turned off.
RAS	(1) Remote access server. (2) H.323 registration, admission and status.
Reactance	The opposition offered to the flow of alternating current by inductance or capacitance of a component or circuit.
Repeater	An amplifier which regenerates the signal and thus expands the network.
Resistance	The ratio of voltage to electrical current for a given circuit measured in ohms.
Response time	The elapsed time between the generation of the last character of a message at a terminal and the receipt of the first character of the reply. It includes terminal delay and network delay.
RF	Radio frequency.
RFI	Radio frequency interference.
Ring	Network topology commonly used for interconnection of communities of digital devices distributed over a localized area, e.g. a factory or office block. Each device is connected to its nearest neighbors until all the devices are connected in a closed loop or ring. Data are transmitted in one direction only. As each message circulates around the ring, it is read by each device connected in the ring.
RIP	Routing information protocol. One of the older interior gateway (routing) protocols.
Rise time	The time required for a waveform to reach a specified value from some smaller value.
RMS	Root mean square.
ROM	Read only memory. Computer memory in which data can be routinely read but written to only once using special means when the ROM is manufactured. A ROM is used for storing data or programs on a permanent basis.
Router	A linking device between network segments which operates at layer 3 of the ISO/OSI reference model.
RSVP	Resource reSerVation set-up protocol.

RTP	Real time protocol.
RTCP	Real time control protocol.
RTSP	Real time streaming protocol.

S

SAA	Standards Association of Australia.
SAP	Service access point.
SDH	Synchronous digital hierarchy.
SDLC	Synchronous data link control. IBM standard protocol superseding the bisynchronous standard.
SDM	Space division multiplexing.
SDP	Session description protocol.
SDSL	Symmetrical digital subscriber line.
Serial transmission	The most common transmission mode in which information bits are sent sequentially on a single data channel.
Session layer	Layer 5 of the ISO/OSI reference model, concerned with the establishment of a logical connection between two application entities and with controlling the dialogue (message exchange) between them.
Simplex transmissions	Data transmission in one direction only.
Slew rate	This is defined as the rate at which the voltage changes from one value to another.
SMDS	Switched multimegabit data service.
SMS	Short message service.
SNA	Systems network architecture.
SONET	Synchronous optical network. Allows low order bit-rate signals to be dropped and inserted without needing electrical demultiplexing.
Standing wave ratio	The ratio of the maximum to minimum voltage (or current) on a transmission line at least a quarter-wavelength long.
Star	A type of network topology in which there is a central node that performs all switching (and hence routing) functions.
STM-1	Synchronous transport module level 1. 155.52 Mbps. Lowest level of SDH hierarchy, corresponds to SONET OC-3.
STP	Shielded twisted pair.
STS-1	Synchronous transport signal level 1. 51.84 Mbps electrical format of the OC-1 SONET signal.
SVC	Switched virtual circuit.
Switch	(1) A device such as a telephone exchange for connecting one circuit to another, (2) A linking device between network segments which operates at layer 2 or 3 of the ISO/OSI reference model. In a layer 2 switch, each port functions as a bridge.
Switched line	A communication link for which the physical path may vary with each usage, such as the public telephone network.
Synchronization	The coordination of the activities of several circuit elements.
Synchronous transmission	Transmission in which data bits are sent at a fixed rate, with the transmitter and receiver synchronized. Synchronized transmission eliminates the need for start and stop bits.

T

T-1	Digital circuit operating at 1.544 Mbps. Corresponds to 24-channels in the North American digital hierarchy.
TAPI	Telephony application programing interface (Microsoft).
TCM	Trellis coded modulation.
TCP	Transmission control protocol.
TDM	Time division multiplexing.
TDMA	Time division multiple access.
TDR	Time domain reflectometer. This testing device enables the reflections user to determine cable quality with providing information and distance to cable defects.
Telegram	In general a data block which is transmitted on the network. Usually comprises address, information and check characters.
Temperature rating	The maximum and minimum temperature at which an insulating material may be used in continuous operation without loss of its basic properties.
TETRA	Terrestrial trunked radio.
TIA	Telecommunications Industry Association.
Time sharing	A method of computer operation that allows several interactive terminals to use one computer.
Token ring	Collision free, deterministic bus access method as per IEEE 802.5 ring topology.
Topology	Physical configuration of network nodes, e.g. bus, ring, star, tree.
Transceiver	(1) Transmitter/receiver. (1) Network access point for IEEE 803.2 networks.
Transient	An abrupt change in voltage of short duration.
Transmission line	One or more conductors used to convey electrical energy from one point to another.
Transport layer	Layer 4 of the ISO/OSI reference model, concerned with providing a network independent reliable message interchange service to the application oriented layers (layers 5 through 7).
TSAPI	Telephone services application programing interface (Novell).
Twisted pair	A data transmission medium consisting of two insulated copper wires twisted together. This improves its immunity to interference from nearby electrical sources that may corrupt the transmitted signal.

U

UDP	User datagram protocol.
UMS	Unified messaging system.
UMTS	Universal mobile telephone standard. The new third generation global mobile communications standard.
Unbalanced circuit	A transmission line in which voltages on the two conductors are unequal with respect to ground e.g. a coaxial cable.
UTP	Unshielded twisted pair.

V

VDSL	Very high speed DSL. Customer access technology providing up to 55 Mbps over 100 m of copper loop.
Velocity of propagation	The speed of an electrical signal down a length of cable compared to speed in free space expressed as a percentage.
VFD	Virtual field device. A software image of a field device describing the objects supplied by it (e.g. measured data, events, status, etc.) and which can be accessed by another network.
VHF	Very high frequency.
VLAN	Virtual local area network. Uses 802.1p/Q protocol with switches to associate nodes in different areas into one virtual network.
VLR	Visitor location register.
Volatile memory	An electronic storage medium that loses all data when power is removed.
Voltage rating	The highest voltage that may be continuously applied to a wire in conformance with standards of specifications.
VPN	Virtual private network.
VSD	Variable speed drive.
VSWR	Voltage standing wave ratio.
VT	Virtual terminal.

W

WAN	Wide area network.
WAP	Wireless application protocol.
WDM	Wavelength division multiplexing.
Wideband	Typically frequencies above 1.5 Mpbs.
WLAN	Wireless LAN.
WLL	Wireless local loop.
Word	The standard number of bits that a processor or memory manipulates at one time. Typically, a word has 16 bits.

X

xDSL	A generic name for a digital subscriber line, a high speed customer access system operating over existing twisted-pair telephone cable.
X.21	CCITT standard governing interface between DTE and DCE devices for synchronous operation on public data networks.
X.25	CCITT standard governing interface between DTE and DCE device for terminals operating in the packet mode on public data networks.
X.25 PAD	A device that permits communication between non-X.25 devices and the devices in an X.25 network.
X.3/X.28/X.29	A set of internationally agreed standard protocols defined to allow a character-oriented device, such as a visual display terminal, to be connected to a packet-switched data network.

Index

THIS BOOK WAS DEVELOPED BY IDC TECHNOLOGIES

WHO ARE WE?

IDC Technologies is internationally acknowledged as the premier provider of practical, technical training for engineers and technicians.

We specialise in the fields of electrical systems, industrial data communications, telecommunications, automation & control, mechanical engineering, chemical and civil engineering, and are continually adding to our portfolio of over 60 different workshops. Our instructors are highly respected in their fields of expertise and in the last ten years have trained over 50,000 engineers, scientists and technicians.

With offices conveniently located worldwide, IDC Technologies has an enthusiastic team of professional engineers, technicians and support staff who are committed to providing the highest quality of training and consultancy.

TECHNICAL WORKSHOPS

TRAINING THAT WORKS

We deliver engineering and technology training that will maximise your business goals. In today's competitive environment, you require training that will help you and your organisation to achieve its goals and produce a large return on investment. With our "Training that Works" objective you and your organisation will:

- Get job-related skills that you need to achieve your business goals
- Improve the operation and design of your equipment and plant
- Improve your troubleshooting abilities
- Sharpen your competitive edge
- Boost morale and retain valuable staff
- Save time and money

EXPERT INSTRUCTORS

We search the world for good quality instructors who have three outstanding attributes:

1. Expert knowledge and experience – of the course topic
2. Superb training abilities – to ensure the know-how is transferred effectively and quickly to you in a practical hands-on way
3. Listening skills – they listen carefully to the needs of the participants and want to ensure that you benefit from the experience

Each and every instructor is evaluated by the delegates and we assess the presentation after each class to ensure that the instructor stays on track in presenting outstanding courses.

HANDS-ON APPROACH TO TRAINING

All IDC Technologies workshops include practical, hands-on sessions where the delegates are given the opportunity to apply in practice the theory they have learnt.

REFERENCE MATERIALS

A fully illustrated workshop book with hundreds of pages of tables, charts, figures and handy hints, plus considerable reference material is provided FREE of charge to each delegate.

ACCREDITATION AND CONTINUING EDUCATION

Satisfactory completion of all IDC workshops satisfies the requirements of the International Association for Continuing Education and Training for the award of 1.4 Continuing Education Units.

IDC workshops also satisfy criteria for Continuing Professional Development according to the requirements of the Institution of Electrical Engineers and Institution of Measurement and Control in the UK, Institution of Engineers in Australia, Institution of Engineers New Zealand, and others.

CERTIFICATE OF ATTENDANCE

Each delegate receives a Certificate of Attendance documenting their experience.

100% MONEY BACK GUARANTEE

IDC Technologies' engineers have put considerable time and experience into ensuring that you gain maximum value from each workshop. If by lunch time of the first day you decide that the workshop is not appropriate for your requirements, please let us know so that we can arrange a 100% refund of your fee.

ONSITE WORKSHOPS

All IDC Technologies Training Workshops are available on an on-site basis, presented at the venue of your choice, saving delegates travel time and expenses, thus providing your company with even greater savings.

OFFICE LOCATIONS

AUSTRALIA · CANADA · IRELAND · NEW ZEALAND · SINGAPORE · SOUTH AFRICA · UNITED KINGDOM · UNITED STATES

idc@idc-online.com • www.idc-online.com

Visit our Website for FREE Pocket Guides

IDC Technologies produce a set of 4 Pocket Guides used by thousands of engineers and technicians worldwide.

Vol. 1 - ELECTRONICS
Vol. 2 - ELECTRICAL
Vol. 3 - COMMUNICATIONS
Vol. 4 - INSTRUMENTATION

To download a FREE copy of these internationally best selling pocket guides go to:
www.idc-online.com/freedownload/